William C. Leone, D.Sc., Carnegie Institute of Technology, is Vice President and General Manager of the Remex Electronics division of Ex-Cell-O Corporation. Dr. Leone has long been active in the management, engineering, and research phases of complex military and industrial electronics and electromechanical systems. He is a licensed Professional Engineer in the states of Pennsylvania and California and holds patents in the fields of machine tools, numerical control, and data processing systems. Dr. Leone was formerly an Assistant Professor of Engineering at the Carnegie Institute of Technology.

Production Automation and Numerical Control

William C. Leone

REMEX ELECTRONICS

THE RONALD PRESS COMPANY • NEW YORK

Copyright © 1967 by

THE RONALD PRESS COMPANY

Library of Congress Catalog Card Number: 67–21679

PRINTED IN THE UNITED STATES OF AMERICA

Preface

This book is intended as a concise survey of modern numerical control techniques and equipment. Its fundamental purpose is to provide management with an awareness of the rapid advances made by automation in the field of production and particularly in the area of automatic, tape-controlled machine tools (to which the generic name *numerical control* has been applied). The implications of the decision to adopt numerical control extend far beyond the immediate questions of shorter production times and reduced work force, and affect such fundamental considerations as plant layout, raw and finished inventory levels, and even the characteristics and variations of the product that reaches the marketplace. The proper and effective utilization of this new technology affects almost every aspect of the industrial establishment. Consequently, numerical control is not the concern merely of the industrial engineer or production manager involved with machine tools, but it is a central issue for all managerial functions—policy making, financial, personnel, and marketing.

The emphasis in this book is upon general technological and economic characteristics of numerical control systems and their components. Comparisons with conventional machining systems are made throughout the text, especially with regard to the cost and production factors involved in evaluation studies.

Although the author has intentionally avoided detailed engineering discussions of system hardware, numerous examples of actual machine installations are given. For example, specifications and comments on the applications of particular machining centers are included to demonstrate the versatility and diversity of equipment presently available and in use.

The decision to convert to numerical control involves major economic considerations. Thus, an example of an actual cost-justification analysis involving the replacement of six conventional machine tools by a single numerical control machining

center has been included and should provide important guidelines for those contemplating similar studies.

The author wishes to acknowledge gratefully the assistance
of the many companies that have generously contributed data
and photographs for use in this book. Specific credit is given to
each in the text.

WILLIAM C. LEONE

Palos Verdes Estates, California
May, 1967

Contents

II. Precision Grinding of Cams. Using Numerical Control with Separate Machining Operations. Machining Center Quality Control. Reduced Handling of Large Workpieces. Machining Versatility with Numerical Control. Versatility of Single-Spindle Universal Tool. Use of Tape-Controlled Rotary Work Table. Tape-Controlled Turret Machining. Turret Drilling: Cost per Part Comparison.

World Trends: Need for Productivity. Industry Trends: Need for Flexibility.

Production Automation
and
Numerical Control

1

The Automation Concept

CHARACTERISTICS OF AUTOMATION

The variations in approach and the range of sophistication levels in applications of automation are so broad that precisely what automation is remains quite elusive. The concept defies simple definition. Nevertheless, numerous authorities have attempted it. As Delmar S. Harder,[1] the Ford Motor Company executive who is associated with the origins of the word, first defined it, "Automation is the automatic handling of parts between progressive production methods from the standpoint that we must consider the effects of automatic handling on each and every phase of our manufacturing processes."

A man who has done much to popularize the word, John Diebold,[2] has referred to the recognition that automation has received as a new set of concepts rather than as an extension of mechanization. He initially thought of it as "a basic change in production philosophy . . . a means of organizing or controlling production processes to achieve optimum use of all production resources—mechanical, material and human." He later testified in a paper submitted to a Joint Economic Committee of the Congress of the United States, "Automation is a philosophy of technology—a set of concepts. In itself, it only

[1] Delmar S. Harder, "Automation, a Modern Industrial Development," *Automation*, August, 1954, pp. 46–54.
[2] John Diebold, *New Views on Automation*, collected papers submitted to Joint Economic Committee, Congress of the United States, 1960, pp. 79–143.

3

makes available to us the knowledge of how to better satisfy our material and intellectual desires."

Roger Bolz[3] wrote, "Automation, the modern name for the marvelous composite of automatic operations and the automatic plant as developed and envisioned, has provoked much thinking and conjecture concerning some of its characteristics and influences. Basically it consists of relating, co-ordinating and integrating machines, mechanisms, controls, and processes to the extent that partial or complete evolution from the primary state to useful end product or service is accomplished automatically, without use of human hands."

According to Joseph Harrington,[4] "Automation is just one link in a long chain—the modern phase of an infinite continuum called mechanization. It extends backward through many centuries to the dawn of civilization, and it will extend ahead through even more centuries."

The definition of Walter Buckingham[5] is that it is "any continuous and integrated operation of a production system that uses electronic or other equipment to regulate and co-ordinate the quantity and quality of production. In its broadest usage it includes both the manufacturing and administrative processes of a firm. These processes can be distilled into four fundamental principles; mechanization, continuous process, automatic control, and rationalization. Each of these four elements has evolved separately. The novelty of automation as a distinct technology is that it is a synthesis of all four, emerging since World War II from a unique combination of scientific breakthroughs and economic conditions."

Sir Leon Bagrit,[6] in the BBC Reith lectures of 1964, said, "Automation is that part of what I have called the 'extension of man' which integrates all sensing, thinking and decision-making elements. . . . It is a concept through which a machine-

---

[3] Roger Bolz, "The New Automation," *Automation,* September, 1954, pp. 14–16.

[4] Joseph Harrington, "A Look into Tomorrow," *New Views on Automation,* collected papers submitted to Joint Economic Committee, Congress of the United States, 1960, pp. 37–43.

[5] Walter Buckingham, *Automation, Its Impact on Business and People* (New York: The New American Library of World Literature, Inc., 1961), p. 15.

[6] Sir Leon Bagrit, *The Age of Automation* (New York: The New American Library of World Literature, Inc., 1965), pp. 35–36.

system is caused to operate with maximum efficiency by means of adequate measurements, observation and control of its behavior. It involves a detailed and continuous knowledge of the functioning of the system, so that the best corrective actions can be applied immediately they become necessary."

Collectively, the foregoing definitions do reveal something about the nature of automation. Certainly, efficient systems are not readily implemented unless their designers have intimate knowledge of the inner workings of the process to be automated. Many disciplines are involved. The technology of automation crosses many lines normally bounding the areas in which the mechanical, electrical, industrial, chemical, and metallurgical engineers, the physicists, the mathematicians, the economists, the social scientists, and the managers operate.

Characteristics of the applications themselves are repeated often enough to suggest a number of generalizations about automation systems:

1. Flow of material is always involved. "Material" here is used in the broadest sense. It may be a bulk substance, discrete products, data, living beings, or even energy. Incidentally, this does not mean that a system must be physically dynamic to qualify as an automated one. The specific function of many systems is to maintain stable conditions in a process. However, this is most often accomplished through regulation, with its concomitant flow of information between elements of the system.

2. Machines are always involved. Again, the word "machines" is used in the broad sense. Their function may be to contain, transfer, alter, or otherwise act upon the material.

3. Control is always involved. This may be of the most mundane type—elementary regulation of a process or simple communication among machines or between a machine and humans. Nevertheless, there is always some kind of interrelationship among elements of the system or between the system and the outside world. The hierarchy that maintains these interrelationships, whether they are inherent in their natural behavior or preordained by a man-made program, is the control.

4. Input and output means are always involved. These provide the interfaces going into and out of the machines in the system. They are the physical contact and communication with

the outside world. They may take the form of data-carrying media (such as punched cards, magnetic tape, punched tape, etc.), measurement apparatus, monitors, displays, printout devices, etc,

In the popular terminology, we can think of the input-output means, the machines, and the controls of automation systems as analogous to the senses, the muscles, and the brain and nervous system of man. Just as all of the latter must be operative to some measure in any activity of man, all of the former are involved in any system that is considered to be automated.

THE BRAINS, SENSES, AND MUSCLES OF AUTOMATION

Perhaps the most advanced applications of automation to date have been in military weapons engineering. There are many examples in fire control, guided missiles, and large complex systems of strategic defense establishments. The level of sophistication in some of these systems is of a very high order. They normally involve complicated equipment that must be operated—often by basically non-technical people—to perform as a part of intricate and extensive systems of human organization, machines, methods, and procedures to accomplish an overall operational result. In each case, the design of the equipment is intimately associated with, and affected by, the surrounding complex of men and procedures.

That the military should be advanced in automation is attributable to its historic pioneering role in this area and to the fact that military programs have made possible the integrated systems approach to problems, an approach which has been labeled "operations research." As described by Dean Wooldridge,[7] "Operations research is a term used to describe the activities of scientifically trained teams who undertake the assignment of analyzing broad areas of operation of business or industrial establishments, to determine procedures and methods of making decision that are most effectively geared to the basic objectives of the business."

As the definition suggests, the principle is not tied exclusively to the military. True, the military was the first to provide the

[7] Dean E. Wooldridge, "The Future of Automation," in *Automation in Business and Industry,* collection of lectures at University of California at Los Angeles, n.d.

proper climate. But the idea of applying a systems approach to assure compatibility of any new development with the overall requirements of the system is also basic to optimum utilization of automation in business and industry.

How about industry? Where is automation being adopted? We know that automation is in use to machine automobile parts, print newspapers, mix chemicals, bake bread, and sort and grade everything from oranges to bank checks. The question is really "How universally is automation being utilized?" and "Are we going fast enough?"

Three major categories of equipment constitute the backbone of all automation activity. These are computers, instruments, and machines. These three are broadly associated with the "brains," "senses," and "muscles," respectively, of industrial automation.

Computers

Computers are also known as "automatic information processors." They may be thought of as encompassing three subfields:

1. Business data handling
2. Engineering and scientific computation
3. Automatic process control

Business data handling is defined to mean those activities which typically involve the gathering of alphabetic or numerical information; inserting this information into file records; sorting, matching, or otherwise rearranging the information; performing computations; and printing or otherwise displaying the resultant alpha-numeric information. Business data handling is characterized by a large volume of input and/or output information, large files, and relatively straightforward computations. Typical examples are payroll, accounts receivable, accounts payable, statistical analysis, and ticket reservations. This field would also include material projection, purchasing, inventory, cost distribution, factory scheduling, machine and manpower utilization—in short, many of the functions generally included in what is sometimes called "manufacturing control."

Systems for engineering and scientific computation are characterized by low volumes of inputs and outputs, relatively small files (low access rate memories), complex computations requiring high computing speeds, and a large high access rate memory for storing intermediate results.

By automatic process control is implied those processes in which measurements are made on material objects, computations are made on those measurements (the computations are frequently quite simple), and the results of the computations are used to control, automatically, continuous or discrete processes that mechanically or chemically affect the observed objects. Typical examples might be in continuous chemical process control such as occurs in oil refineries or in the processing of photographic film emulsions, or in piecewise processes such as automatic tool control or material movement. Other examples are automatic inspection and automatic flight control.

Computers are daily being applied to a greater variety of tasks for the simple reason that they can handle certain jobs better and more efficiently than humans. In particular, a computer can cope with more data and can process it much faster than man. It is these factors, speed and capacity, and not a nonexistent "thinking ability," that make the computer so useful.

The computer is increasingly becoming involved in every facet of our lives. Directly or indirectly, a computer is associated with almost every statistic with which we are identified, every printed word we read, and every product we use. The myriad ways by which it enters our lives are almost unbelievable, yet they are expanding daily.

Paperwork.   Computers are increasingly being used where a lot of paperwork has to be handled. Paperwork for an enormous amount of recording and communications is basic to all business. Computers handle and process millions and millions of words and numbers in the day-to-day routine of most businesses. They are vital to the greater part of transactions involved in the accounting departments of many companies, including payroll, accounts receivable, accounts payable, invoicing, tax accounting, etc. The purchasing functions would be lost without the help of computers in stock control, inventory records of all kinds, shortage accounting, etc. Manufacturing activities increasingly involve computers for production control, testing, and quality control. Personnel and marketing departments are putting their records on media amenable to processing by computers for numerous analyses to assist them in their respective functions. Last, but certainly not least, general managers are becoming more and more conscious of the fact that computers can keep them current on vital operating data far beyond what was available to them before.

What is outstanding about this widespread use of computers is not only its inherent labor-saving characteristic; certainly the computer plays the role of a superefficient clerk. It is also that its speed and capacity are such that it handles more data, faster, than would have been possible with the entire world's human effort applied to the task if previous limitations of manual capacity only were applied. Moreover, the computer can perform far beyond merely processing data fed into it. It is useful in monitoring and logging data as they are created at rates not remotely possible before its advent. This makes the computer useful in processing and taking action on information before humans are even aware of its existence.

The computer's capability in summarizing is perhaps the characteristic that is most useful in the preparation of paperwork reaching management. In centralized or integrated data processing installations, the computer is increasingly given the task of extracting segments of data processed and summarized in a variety of ways to present to management the totals it needs to make decisions. This carries with it the obvious implications of the computer's ability to act as a "tickler" in alerting management to trouble before it happens by "red-flagging" at predetermined summary mileposts. This capability of a computer points up the advantages of its non-thinking characteristic. It takes nothing for granted. Once it is programmed to "red-flag" a prespecified trouble spot, it does so without prejudging the merits of an alarm.

Industrial Processes.   Many industrial processes require the simultaneous collection and processing of data from literally hundreds of sources. Often the sources are widely scattered or remote from the data processing center. Each of these sources may carry hundreds, thousands, or even millions of data units every second. To classify these data and to route them to the proper inputs of a computer are themselves jobs for another computer.

In some of these processes, computers retrieve and evaluate signals too well hidden in noise to be detected by a human operator at all. These cases require the computer to furnish accurate solutions or evaluations quickly, either for a fixed condition or for changing situations. A military counterpart to this type of industrial process is the long-range missile launch, which involves the correlation of information from many missile ranges and tracking stations to assure its correct performance.

Computers are often used to control industrial processes which consist of a complex machine system responding to data externally fed into the system. They may be either an integral part of the system or general computers whose output in part is utilized to control a particular system.

Design.   A characteristic of many engineering design problems is that they are iterative in nature, that is, require a long series of repetitive arithmetic computations for solution. Many of these techniques require a great deal of effort and time when applied manually, and, therefore, they were not sufficiently utilized until the computer made computation possible at tremendously high speeds. Computers are being used more and more in design where the mathematical manipulations would otherwise be so tedious and lengthy as to preclude adequate consideration of many available choices or approaches.

Training.   It is often expensive, dangerous, and impracticable to train a large group of men under actual conditions to operate an aircraft or an atomic reactor, control a satellite, or operate a space ship. A computer can simulate the operational conditions for a trainee, react to his actions, and show him the results of his actions. Thus, the computer lets him get many days of on-the-job experience without personal risk and without risking the destruction of expensive equipment.

Instruments

With the development of science and technology, the capacity of machines increased beyond the point where man's muscle power was adequate to maintain control. This brought about the creation of systems for coordination of operation of units without direct human participation. However, this pointed up man's inadequacies in observing and sensing the characteristics of the process in operation so that devices had to be developed to remove man also from the functions of monitoring the process. In achieving automation, then, the aim is to leave man with the functions only of programming the operation of machines, setting up, starting, and stopping machines, and supervising the operation.

Monitoring and Control Instruments. The requirement for monitoring devices is just one indication of automation's inseparable link with instrumentation. A characteristic of automation systems is response of interrelated subsystems and elements, so instruments for measurement, indication, and control are vital to automation progress. Instruments for the automatic control and regulation of technological processes that are part of this progress are those for the measurement of temperatures, pressures, flow rates, liquid levels, and displacements; instruments for the determination of the composition and properties of materials, for the determination of viscosity and density of materials, and for the determination of concentrations of solutions; weighing devices; magnetic and optical instruments; and instruments for measuring many other physical parameters.

Automatic instruments help solve control problems in the operation of industrial processes by measuring specific controlled variables and setting in motion operations for maintaining them at predetermined values.

Trends in Instrumentation. Development of instrumentation has been rapid indeed in recent years. The various physical parameters with which industry is concerned in applying automation can be measured by devices (or transducers) that are very precise and highly reliable. Some progress has been made in designing the external characteristics of these devices so that their interfaces with other equipment with which they must work are somewhat standardized. For example, if a transducer is to sense a physical parameter and translate its measured value into one that can serve as an input to a system, it is most desirable that the sensing part of the transducer indeed "see" the physical parameter at its true quantitative level. It is also important that its output be in a form readily acceptable by the system that will use the data and process them.

Much is still to be done in this area, however; the instruments and devices that do the sensing, monitoring, and indicating represent the interface of any automation system with the outside world—the physical parameters which comprise the inputs to the process to be controlled. Consequently, they can "make or break" an automation installation, depending upon the reliability of the data they report. The precision, reliability, range, response, and speed of many automation systems are limited by the instruments and devices available to do the sensing, monitoring, and indicating.

Machines

The machines in automation systems represent the "muscles" in those systems. Without question, the most important industry associated with machines is metalworking. It is interesting to trace briefly some of the trends in the art of metalworking leading to its present state. Of course, the development of these trends did not always occur sequentially, but, rather, many developments took place concurrently.

Replacing Muscle Power.   Certainly, one of the first notable advances in production came in the step from hand tools to machines which made it possible to substitute for human motive power. A natural succeeding step was to make the power available to a number of machines. This brought about the metalworking shop described by Colvin,[8] with "power by a system of belts and line shafts from a Carliss steam engine (25 hp) in the basement for 34 machines." As pointed out by Ashburn,[9] standards for fits, threads, tapers, and a thousand elements of both machine and product design had to be developed.

The Production Line.   While standards progressed from the micrometer to the master gage block to somewhat automatic accurate measuring equipment, interchangeability of parts became practical and the continuous production line was made possible. This represented a significant change from the "batch" method of production, where, as Ashburn described it, "a batch of parts would go to one spot for turning, then to another department for drilling, then across town to be hardened, then back to a third department for grinding. When we had all the

[8] F. H. Colvin, *60 Years with Men and Machines* (New York: McGraw-Hill Book Co., Inc., 1947).
[9] Anderson Ashburn, "The Development of Automation in Metalworking" (New York: The American Society of Mechanical Engineering, 1955). Preprint.

parts made, we'd take them across the street and put them together—coming back to fix any that wouldn't fit."

The demands of the production line became more pronounced as the degree of specialization in preliminary operations increased, so that the natural result was the combining of two or more operations on a single machine bed. Thus, there was a trend from individual machines to transfer machines. As material handling showed itself as an obstacle in accomplishing high production rates, the transfer machines were linked into automatic production lines and more complex forms of integrated processing evolved.

It can be seen that the aforementioned trends led to the "Detroit" type of automated mass production, where very large quantities of a given product are manufactured at high speeds and low per-unit costs. This usually requires that the product approach be used; that is, highly specialized machines and production lines are designed, made, and arranged for the most efficient high-speed manufacture of a particular product. The great majority of the men operating these machines do not necessarily require the general talents of machinists but, rather, are true "machine operators" whose skills are limited to given routines. Special-purpose machines have been highly developed in the United States during the last two or three decades and applied with spectacular success in large-volume industries. The large domestic market for the mass-produced articles in the United States has justified the use of these high-speed machines. The economies brought about by their use on long production runs have justified their relatively high capital costs and have proved them to be a sound investment. These economies include not only the direct ones, that is, lower floor-to-floor times for a given operation, but also those reflected indirectly in a smooth production flow, efficient utilization of factory floor area, reduced labor requirements, and substitution of lower-skilled labor than that required for operating conventional universal tools.

Systems Approach to Production.   In recent years, there has been an effort to add flexibility to cope with the decided trend to greater diversity in configuration and quantity demands. This effort is meeting with varying success. We are seeing tape-controlled machines, production aids to operators, position displays, materials handlers, conveyors, machining centers, and the like. There is no question that great strides are being made.

Unfortunately, too often each attempt gives a solution valid only for that part of the problem being considered at the time. The production of an item must be given the approach mentioned earlier. It must be treated as a system from the first input to the final product. The input, the data with which one starts the fabrication of a part, must be evaluated. All too often, blueprints utilizing conventional techniques are used for this representation. There is considerable expense involved in this step. When one considers that the real function of the blueprint has been to tell the machinist (and assembler) indirectly how he must operate his machine to produce the part described, the question arises as to whether there might not be a better way to present the original data. These data may often be put in a form more suitable for consumption by the machine intelligence that has replaced the human intelligence formerly used.

There is now a realization that machines and their controls must be an entity and that the predominant situation of the recent past, when machines and controls were conceived and developed separately and only later integrated as a system, is not good. Both machines and controls must be made up of elements that can be integrated for as complex a set of functions as the output warrants, yet be modified for other functions with relative ease. The desirable result is a variable but basic machine for general-purpose use. A great many operations, even within the mass-production industries, can be performed just as rapidly by a general-purpose machine, designed for what may be termed "average" conditions, as by one of more complex design. This can best be done if all parts of the job—machine and controls—are designed together as one system.

For most shops, automatic production (especially for small lots) suggests a realignment of the work force. Since these same shops will concurrently have conventional manufacturing operations, at the very beginning of the job the part must be evaluated to determine whether it is to be made by the automated equipment or by more conventional means. This procedure can be, and has been, routinized in many companies. However, it can be the root of serious problems in cases where equipment-application decisions are made too casually. Automation suggests changes not only in equipment but also in management procedures to optimize their utilization.

APPLICATIONS OF AUTOMATION

There are few industries today that are not automated to some degree in their operations. Specific examples of existing automation systems are numerous; many more are in the planning stage. The purposes served cover the full range of commercial and industrial activity.

Industries Adopting Automation

The wide variety of installations makes it difficult to define a typical system. However, there are typical needs met within particular industries.

Service Industries. Utility companies have a surprisingly high proportion of their staffs engaged in clerical activities. As a rule, they are large organizations dealing with a large majority of the public in their areas. This means substantial benefits are derived from economical information processing made possible with automation.

Governments, whether local or federal, may be considered basically as vast service organizations. Information for government agencies such as the Census Bureau is sometimes gained through field workers' making out handwritten reports and forwarding these to central offices. It is also often gained through mandatory reports to various agencies by individuals and industries. Most frequently, the primary problem is the processing of vast amounts of data to obtain a much smaller amount of significant data and to determine reliability of field reports. A large number of correlations must be determined, and a large number of tables must be made up from the data obtained.

Government agencies were among the first to spur activity in automatic information processing, because of the tremendous amounts of data which must be obtained and correlated from so many sources, including extensive programs in taxation, social security, and unemployment compensation, and other insurance programs and procurement. In addition, there are the many armed-forces programs, which include requirements for the most extensive and exotic systems.

Process Industries.   Labor cost in comparison with product value in the petroleum industry is likely to be misleading. Although the proportion of total product cost spent on labor is among the lowest of all industries, there still remains a high dollar volume of clerical work performed in the oil industry. In addition, there is the problem of finding the best combination of end products to produce from a given amount of crude oil. The solution of this problem presents a possible saving or added revenue to the industry in excess of complete saving of all labor costs.

The same problem exists for chemical process industries as for refining processes within the petroleum industry. Because they are much alike, controls must be exerted on the various processes within the limits of the prescribed operational routine. In addition, there often exists a choice of the most economic combination of products which can be produced from various quantities of raw material. Information for making the product-mix decision is gathered from automatic measuring devices and from cost data obtained from accounting records.

Merchandising.   Department stores use systems which assist them in certain of their clerical operations. The main problems encountered in department stores are in point-of-sale recording and inventory control.

Finance.   Banks and other financial institutions represent an industry where nearly all of the activities are devoted to clerical work and its related functions. Among the major industries, insurance companies rank second only to banks and finance companies in the amount of effort expended on clerical work. In addition to the savings that accrue by the reduction of this labor force, risk is greatly reduced and additional savings are brought about through the decrease of the time lag between the collection of data, their processing, and the availability of the data to management.

Manufacturing.   The amount of clerical work involved in product manufacturing is much larger than is ordinarily assumed. In addition, of the major industries, product manufacturing probably has the highest proportion of operating control reports prepared by the lower levels of its organization. These include the following:

1. *Scheduling.* Certainly one of the most important factors in profitable operation in manufacturing is that of scheduling man and machine utilization as efficiently as possible, while holding inventories to the most economical levels. Data are entered from estimated operation times. The processing is much more complex than can be assumed in the case of many other accounting problems, because of the vast amount of variable data.

2. *Material Control.* Accurate records of all material must be kept for tax purposes, determination of financing advantages, manufacturing schedules, procurement schedules, economic purchasing, and control of pilfering and misuse. Information is entered into the system from a wide variety of locations by many different people on many different forms. Many of the items which seemingly have inconsequential value and, therefore, may not warrant expensive paperwork techniques can be, if uncontrolled, bottlenecks in production.

3. *Product Cost.* Cost records are kept on various products in order to determine their profitability, to decide whether to "make or buy," and to establish inventory values. Whether the costs are maintained on a job cost basis or a standard cost basis, the manipulations involved in ensuring that the costs are current in a system where the turnover is large and the product ingredients or models are many are exceedingly complex. Automatic systems provide a visibility to management on true costs previously not available.

Industrial Applications

Process-control automation is prevalent in broad areas in the chemical, petroleum, iron and steel, cement, paper, textile, and food industries, and in utilities, discrete products, machine tools, and services. The installations vary from the most elementary, where simple data logging is done, to multiunit installations, where many or all processes within a plant (or plants) are started, stopped, monitored, and controlled, and measurements associated with the processes are displayed, recorded, and transmitted to other centers.

Logging in the Chemical Field.   Technological advancements in sensors, and instrumentation in general, have provided much of the impetus in making automated process control a reality in the chemical field. Logging systems, for instance, are now available that can automate completely the collection of field data. Loggers in single-channel or multichannel forms can sample inputs from many external sensors. Data in analog or digital form can be collected from transducers for measuring current, voltage, power consumption, gas flow, temperature, pressure, displacement, and many other parameters. The recording can be made on any of various available media—graphs, charts, films, tape, etc.—and can be transferred to magnetic or punched tape, or punched cards, for subsequent processing by a computer. A final typewritten tabulation is made available for use.

Sand Casting.   Automation has invaded the unlikely field of sand casting,[10] perhaps the most difficult metalworking process to automate. The Chambers Division of the Altamie Corporation, Shelbyville, Indiana, has made a foundry system in which mold and core sand are conveyed into precision molding, core setting, mold assembly, metal pouring, cooling, shakeout, and back through a sand-reclamation system to molding again. The system uses the carbon dioxide processes for hardening of molds and cores and utilizes belts and walking-beam conveyors with transfer mechanisms similar to those used with machine tools. A normal 20-man operation for production of 120 molds per hour is accomplished with 4 men—a control operator at the console, and a monitor each for sand mixing, melting, and shakeout—and yet the castings are superior in qualities such as density and surface finish.

Steel Industry.   Practically every known control problem can be found in the modern integrated steel plant, which Schuerger and Slamar[11] described as being made up of a group of massive, complex, interrelated, individual processes. The chemical,

[10] A. W. Young, "Circle of Sand Concept Automates Foundry," *The Iron Age,* February 17, 1966, pp. 120–122.

[11] Thomas R. Schuerger and Frank Slamar, "Control in the Iron and Steel Industry," *Datamation,* February, 1966, pp. 28–32.

metallurgical, and thermal aspects of the steelmaking process
seem to be fairly obvious and simple. In practice, this is not
quite so. Measurements are difficult to make and process.
Theory is complex. Consequently, precise knowledge of the
state of the process is not always available. During the rolling
operations, massive problems in material handling are encountered. Tons of steel must be manipulated on roll tables, often
at extremely high temperatures. With sheet and strip mills,
operating often at several thousand feet per minute, precise
control must be maintained over both temperatures and finished dimensions. It is apparent that the control required in
the modern iron and steel industry is both broad and intensive.
Chemical analysis must be established for raw materials, liquid
iron and steel, and solid steel products. Temperatures of liquid
and solid metals in various states must be measured. Dimensions of semifinished and finished products must be measured.
Environments in the steel industry are harsh, and measurement
and control equipment must be extremely durable to function
properly. Yet, automation systems for steelmaking from blast
furnaces to rolling mills to tinning lines are in being, and more
are on the way.

OPEN HEARTH. A computer-run open-hearth system is in
operation in Tokyo (Kawasaki Iron Works, Nippon Kokan
K.K.) for one of Japan's major steel producers. The open-
hearth furnaces are automatically regulated by a system which
determines the amount of ore needed to attain a fixed melt-
down temperature of the charge, the proper time to add hot
metal, the amount of oxygen needed for melting, the desired
temperature of the furnace, the progress of carbon reduction
during refining, the time to charge ferroalloys, the time of completion of the refining process, the amounts of manganese and
molybdenum needed, whether a carburizing agent is necessary,
and, if so, the amount needed. Reported advantages include
not only better and more uniform quality of product but also
reduced use of additives and increased operational efficiency.

Heat Treatment. Rex Chainbelt, Inc., of Milwaukee, has applied automation to heat-treat furnaces for many different kinds
of parts.[12] Each of them has individual requirements for car-

[12] Milton Gussow, "Automated Heat Treating Furnaces Insure Repeatability
at Rex Chainbelt," *Metalworking News*, February 14, 1966, p. 23.

burizing and straight heat treatment. More than 20 different cycles are processed through the automated line on a regular production process. The furnace does not care or know what is going through it at any one time. By changing of clip cards, exact repeatability and accurate quality control are ensured today, next week, or next year. Also, by exact repetition of processes points of a cycle can be accurately determined and possibly minimized, thereby reducing total processing time for any type of heat treatment or material.

Automobile Industry. No discussion of automation can be complete without the inclusion of remarks on the automobile industry. It is widely conceded that mass production probably had its beginnings at the Ford Motor Company, where the first installation was made of progressive assembly of flywheel magnetos on a moving conveyor. Today, automatic machines drill, mill, bore, ream, broach, hone, tap, inspect, and quality-control engine blocks, cylinder heads, crankshafts, stamped parts, gaskets, and many other parts for automobiles.

Even more dramatic is the advent of machine systems utilizing stereophotogrammetry and digitized drafting techniques to cut the lead time between the designer's clay model and the finished car. With this system, the physical specifications of the clay model are converted into digital information defining its shape. These shapes generally are too complex to be expressed mathematically, since they are part of an aesthetic concept. The digitized design information is processed through a computer, which prepares tapes for numerically controlled machine tools. Production tooling is then made from the tapes. The design data in numerical form are also fed into numerical control drafting machines, and drawings, including perspective views, of the parts and the complete automobile are made.

Manufacturing. The interest in applying automation to eliminate some of the steps between design and the manufactured part has been mentioned. Another step along these lines has been taken by IBM with its Automated Manufacturing Planning Program, which generates detailed manufacturing instructions from engineering designs. Manufacturing is assumed to be an iteration of simple alternative decisions much like a "relay tree." In an application of the system at the Rollway Bearing Company, Syracuse, New York, for instance, the inputs are design specifications for roller bearings and numerous characteristics of materials to be used versus target load ranges, machines to be used versus tolerance requirements, etc. Detailed manufacturing routings are the output.

Numerical control in metalworking is widespread and growing, beyond the well-known cutting applications, in such dissimilar uses as wire wrapping, arc welding, tube bending, testing, shot peening, and many other processes.

Efficient Use of Computers

The computer is lending its versatile services to practically all areas of industrial activity. A problem exists, however, in its misuses by a segment of business and industry that does not thoroughly comprehend it. Much of industry first accepted the computer through accounting. Its use in the payroll function was one of the first obvious applications. The investment is often too great for using the computer for payroll alone, so pressures arise for use in other areas within the company. These other areas quite often are in production or materials control. The variety of work from the financial, manufacturing, and engineering areas presents load and scheduling problems. Immediately this prompts the question of jurisdiction over the data processing facilities. In some companies, jurisdictional disputes over computer priorities have caused a veritable tug-of-war among departments as to whether there should be integrated data processing or not.

Once it is decided to use the computer for a function, say materials control, the format of the data and a particular type of computer equipment must be chosen. It is not always possible at the beginning of computer use within a company to foresee the scope of needs several years hence. Yet the investment is large, and the cost of changes in either data format or equipment type militates against frequent or drastic departures from the initial approach.

The foregoing is not intended to minimize the dramatic progress made in computer usage. However, much still remains to be done in obtaining proper utilization, flexibility, and interchangeability of certain equipment.

2

What Numerical
Control Is

NUMERICAL CONTROL: A FORM OF AUTOMATION

The generalizations about automation systems touched on in
the preceding chapter apply to *numerically controlled systems*. Flow of material exists—both in the tangible form of the
work that is operated upon and passes through the system and
in the form of information which flows into and within the
system. The machine tool and its actuators for drives and
auxiliary machine functions represent the "muscle." Monitoring
devices, reading mechanisms, and data transfer media (such as
tapes or cards) are the "senses." And there is a control unit,
which maintains communications among, and influence over,
the other elements of the system.

INFORMATION FLOW

Although it cannot be thought of as being that restrictive
today, automation was initially most closely associated with
manufacturing and the tools required for it. In fact, until
comparatively recently, most of the integrated steps that comprise manufacturing—raw materials handling and processing,
product manufacturing, inspection and quality control, assem-
bling or mixing, testing, packaging, warehousing, and shipping

—were given little attention in attempts to automate.[1] The process phase of manufacturing—production and, particularly, fabrication as applied to machine tools in metalworking—has been the source of the considerable activity we know as "numerical control" (N/C). This is the name given to the concept that involves the control of automatic equipment for performing a task in accordance with numerical data. The data are fed to the equipment as a series of fairly elementary operations which together make up the task as a single operation.

The processing equipment in numerical control is most often an automatic machine tool, which performs machining functions without operator control or manual handling except for setup and part removal. Coded information, consisting of machine operations and control instructions to the machine, is transmitted to a control unit, which also receives data from monitoring devices and reading mechanisms indicating work and tool positions. From these inputs, the control unit derives signals for controlling the power drives to move the work or tool.

Block Diagram

Figure 2–1 shows the general information flow in a numerically controlled machine-tool system. The process from initial specifications to finished part ordinarily comprises two phases. The first consists of information handling from the initial specifications to a data transfer medium in which the machine instructions are incorporated, that is, dimensional information plus data for sequence and control functions. This is the phase normally associated with the planning or programming department in a conventional manufacturing plant. The second phase is associated with the machine shop and consists of information flow through the machine system itself, starting with the data transfer medium containing machine instructions and resulting in the finished part on the machine.

Programming the Part

There are many possible forms for specifying the part to be fabricated. A sketch with pertinent dimensional information, a

[1] Roger Bolz, *New Views on Automation*, collected papers submitted to Joint Economic Committee, Congress of the United States, 1960, pp. 13–15.

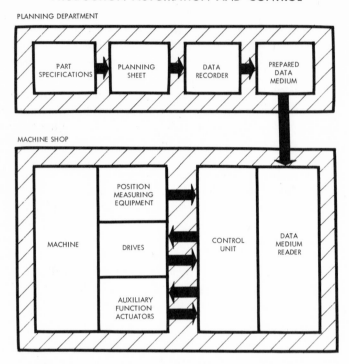

Fig. 2–1. Numerical control system information flow.

blueprint, a set of mathematical equations, or even a duplicate of the part itself may be used to provide the essential input data for generating machine instructions. The part specifications are usually tabulated on a planning or program sheet. This is done to put the appropriate part and machine data in proper form to serve as a sort of checklist and to facilitate subsequent recording for machine use. With most numerically controlled installations in present use, the planning or program sheet is prepared from information given on the standard engineering or production drawing currently used in conventional machining. It shows the vital statistics of the part—material, tools to be used, etc.—plus the operations to be performed and the dimensions to be achieved in their proper sequence. The dimensions or displacements are based generally on a three-dimensional system of rectangular (or Cartesian) coordinates. Figure 2–2 shows a space frame denoting the axis terminology normally used. This is in accordance with the axis and motion nomencla-

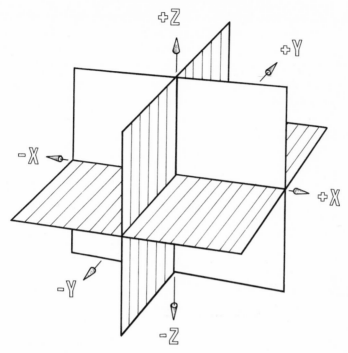

Fig. 2–2. Rectangular coordinate system for dimensions.

ture established by the Electronic Industries Association for numerically controlled machines.

The Z axis of motion is parallel to the principal spindle of the machine. If there are several spindles, one is usually selected as the principal one. The X axis of motion is horizontal and parallel to the workholding surface. The Y axis of motion is perpendicular to both X and Z and is positive in that direction which makes what is known as a right-handed set of coordinates —that is, one in which +X rotated into +Y advances a right-handed screw in the +Z direction. Rules also have been set for complex machines in which there are more than three axes. Although most programming follows the foregoing standards, there is a considerable amount of programming which does not not necessarily conform in every respect.

The additional information that must be provided by the planner is the data corresponding to actual movements he wants the machine to make. Whereas, in conventional machining, a

| PART NO: Sample | TAPE NO: | MATERIAL | PROGRAM & SET UP SHEET FOR M33 | | SHEET 1 OF 1 |
| PROG | CH'K | APPV'D | | COMPANY | |

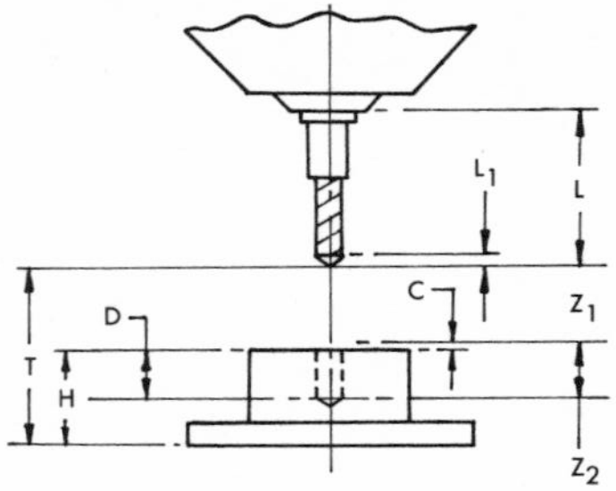

TOOL NO.	TOOL	L	L_1	FEED RATE	SPEED
1	CENTER DRILL	4.000		4.00 "/MIN.	3000 RPM
2	1/4" DRILL	5.000		4.00 "/MIN.	1000 RPM
3	1/2" C'BORE	7.000	00.250	2.50 "/MIN.	625 RPM
4	#7 DRILL	5.000	00.060	4.50 "/MIN.	1500 RPM
5	1/4-20 TAP	5.000		11.00 "/MIN.	225 RPM
6	1/2" 4 FLUTE END MILL	5.000		—	840 RPM
7					
8					

PREPARATORY FUNCTIONS

CODE	
g80	(MILL, TAP OFF) CANCEL CYCLE
g81	RAPID
g82	DWELL
g84	TAP (SPINDLE REVERSE)
g86	MILL ON
g88	RETRACT TO LIMIT

MISCELLANEOUS FUNCTIONS

CODE	
m00	STOP
m08	COOLANT ON
m09	COOLANT OFF
m21	TOOL OFFSET
m23	MISC. AUX. 1
m24	MISC. AUX. 2
m25	MISC. AUX. 3
m26	MISC. AUX. 4

L = PRESET TOOL LENGTH

L_1 = AS SHOWN

H = HEIGHT OF PART & FIXTURE AT POINT OF OPERATION

T = TABLE TO CUTTING TOOL (SET UP OPERATION)

D = DEPTH OF HOLE

C = CLEARANCE

$Z_1 = T - H - C$ (RAPID APPROACH)

$Z_2 = L + D + C$ (FEED)

Fig. 2-3. Setup sheet.

machine operator determines the cuts to be made to achieve certain final dimensions and finishes, with numerical control, the planner is essentially the "machine operator on paper." Data on tool selection, speeds, feeds, and coolant control, if applicable, are established by the planner and not left to the operator's judgment each time the part is run. At this point, the planner also documents information required for machine setup. Just what form this takes depends upon the particular planning sheet used. In some cases, the planning sheet accommodates both tape and setup data. In other cases, a separate setup instruction sheet is prepared for transmittal to the machine area. Figure 2–3 shows a typical setup sheet used by the planner in preparing a program. In this example, it is used to record preset tool lengths and feeds and speeds for each tool that is to be used. Because it is necessary to have certain information about machine setup (such as tool lengths when computing the Z-axis program), this form is completed before the planning sheet. This particular setup sheet includes a diagram illustrating the nomenclature used for the known or specified distances. The figure can also be used as a guide in the programming of the vertical axis.

The planning or program sheet provides a systematic method of presenting the program in order to facilitate the preparation of the tape. Before attempting to program a part, the planner familiarizes himself with the machine capabilities, operating features, and control.

Upon receiving a blueprint, the planner's first job is mentally to plan how the part would be located within the travel capacities of the machine tool. Following general practice, the drawing is usually laid out with decimal values for locations. Because the tape-reading and axis-positioning times are generally much shorter than the tool-changing time (if applicable), as much work with each tool is done as is practical before changing tools. Operations are grouped in sequence according to the normal order of machining, that is, center drilling of all holes, then drilling of all holes, then tapping, etc. Figure 2–4 shows a typical planning sheet for a positioning system. In the column headings, from left to right, are all the words in the order ultimately listed on the tape. Each machining operation, then, is fully described in one line of the planning sheet and is, in turn,

PROGRAM SHEET

PART NO. _____   TAPE NO.: _____   CHG. _____   BY _____   DATE _____   CO. _____   SHEET _____ OF _____

REMARKS	SEQUENCE NUMBER	TAB	PREPARATORY FUNCTION	TAB	DIMENSION X +	TAB	DIMENSION Y +	TAB	DIMENSION Z +	TAB	FEED RATE	TAB	SPINDLE SPEED	TAB	TOOL SELECTION	TAB	MISC. FUNCTION	END OF BLOCK
	n		g		x		y		z		f		s		t		m	
	n		g		x		y		z		f		s		t		m	
	n		g		x		y		z		f		s		t		m	
	n		g		x		y		z		f		s		t		m	
	n		g		x		y		z		f		s		t		m	
	n		g		x		y		z		f		s		t		m	
	n		g		x		y		z		f		s		t		m	
	n		g		x		y		z		f		s		t		m	
	n		g		x		y		z		f		s		t		m	
	n		g		x		y		z		f		s		t		m	
	n		g		x		y		z						s		t		m	

Fig. 2—4. Typical planning sheet.

punched as one block of tape. After the machining pattern and sequence are established, the planner transfers this information to the planning sheet.

The information documented by the planner is of two basic types. There are data that refer to the movements of the slides and data that refer to auxiliary functions. Although both types of data are in numerical form, the significance of the numbers usually differs in each case.

In the former, the numbers correspond directly to the inches (or other units) of slide movement. If the information for each movement is given in terms of desired displacement from the last position, the system used is said to be "incrementally programmed." This means that the data are taken from the drawing in "increments" of movement. On the other hand, the slide-movement information may be given in terms of distances from a particular location. The particular location is known as a "fixed zero reference" and the system used is said to be "absolutely programmed." The absolute type of system is particularly prevalent for two reasons. First, many of the available drawings use base-line referencing in depicting part dimensions. Absolute programming permits one-to-one correspondence of programmed numbers with those on the drawing, without the necessity for intermediate arithmetic computations. Second, many earlier numerical control systems incorporated "absolute" measuring means, that is, the position-measuring equipment did its job in terms of actual location from a fixed reference and the natural inclination was to keep the whole system absolute. However, absolute programming does not have to have absolute position measuring, and, in fact, many systems today utilize absolute programming and incremental position measuring.

It should be noted that several incremental programming systems are available also. There are pros and cons for both absolute and incremental programming, dictated usually by the work to be performed. A job that can be done with one system can be done with the other. However, depending upon the drawing reference system and the nature of repetitive positioning steps required, one system may turn out to be less tedious to apply on certain jobs than the other.

The planner also lists supplementary information including, where applicable, data on speeds of traverse, tool selection, spindle speed, rotary table index position, feed rates, coolant start/stop, etc. These are known as "auxiliary functions" and are characterized by codes which do not have the same numerical significance as in table movement. The auxiliary-function codes, often of the type which signify "yes-no" or "on-off" decisions, allow the planner to make a selection of feeds, speeds, and the like.

Positioning Versus Contouring

Because of the significant differences in approach in both preparation of instructions and equipment required to implement them, comments are in order regarding two basic categories into which numerical controls for machine tools fall, namely, positioning systems and contouring systems. The required movements of the work or tool dictate which of the two categories applies.

In positioning systems, which are also known as point-to-point systems, the operations to be performed can be described in terms of discrete work or tool position coordinates without reference to the movement between these points. If there is movement in more than one axis at a time, coordination of the multiple movements is not ordinarily available or necessary. In most cases, positioning systems are used for moving the tool or work from point to point for actual machining operations (such as drilling, tapping, reaming, etc.) to be done at the terminus after each axis movement has stopped. Positioning systems also permit certain machining operations (such as milling) to be performed during axis movement. However, in these cases, the machining is done either in straight lines parallel to the particular machine axis moved or in a substantially random path between the programmed points. Exceptions to the latter exist where partial coordination of two axes is used for approximating continuous path work with a positioning system. This has definite accuracy and speed limitations and is accomplished essentially through specialized programming rather than true coordination of axes in the equipment.

In contouring systems, which are also known as continuous path systems, the requirements ordinarily are to machine parts in irregular or curved shapes. The work or tool movements must be in a prescribed path continuously under control. This can be accomplished only through coordination of the movements of the various axes. This, of course, points up the main reason that contouring systems are more complicated than positioning systems. Not only is the physical equipment for coordinating the machine drives significantly more sophisticated, but the input information for describing the part is more complex. Every point in the continuous path describing the part shape must be accounted for, whereas, in positioning, only the end points of a movement are explicitly defined.

There are several different approaches for the path description in contouring work. Many types of equipment utilize so-called linear interpolation. This requires that a curve representative of the actual part shape be approximated by many straight lines drawn between successive points with known or calculated coordinates on the curve. For some systems, the straight lines are drawn tangent to the curve at numerous points. Of course, the approximation to the curve becomes better as the number of prescribed points, and therefore straight lines, is increased.

The so-called interpolation approach to curve approximation is not restricted to short, straight lines drawn between successive points. Portions of other curves can be and are used with some systems, and, in fact, some use combinations, such as linear plus circular interpolation. Systems which use other than straight lines are apt to be parabolic interpolation systems. This does not signify that the approximating curve segments are necessarily only parabolic—they may be parts of parabolas, circles, ellipses, or hyperbolas, that is, any curve that can be described by a second-order equation. The path description, then, is accomplished by approximating the curve representative of the actual part shape by numerous short segments of easily definable straight lines or curves.

Contouring systems also have a complexity associated with the fact that provision must be made for the cutter diameter. The movements programmed are generally those for the machine slides to which the cutter is fixed. Therefore, the path delineation is that for the cutter center. Yet the cutter edge is in contact with the part periphery. The cutter center path must be computed to accommodate an offset equal to the cutter ra-

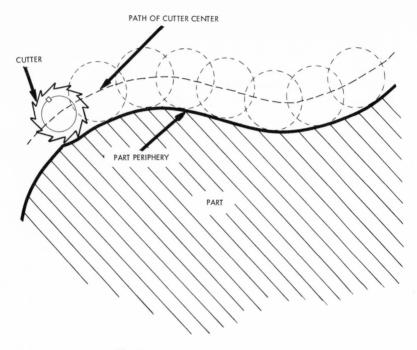

PATH OF CUTTER CENTER

CUTTER

PART PERIPHERY

PART

Fig. 2–5. Cutter offset in contouring.

dius (Fig. 2–5). In any path other than one that assumes the cutter to be of zero diameter, the curve for the cutter center is mathematically different from that for the cutter edge or the part periphery.

Clearly, contouring systems involve a higher order of computation and equipment of corresponding complexity to do it.

Recording Data

After all of the data pertinent to the manufacture of the part have been entered on the planning sheet, they are ready for recording on a data transfer medium. In some cases, a permanent data medium is not used at all. Rather, the data may be entered into dials or switches manually for temporary storage in other memory devices until used by the controls.

In most cases, however, the machine instructions are put on a permanent, readable, and portable medium. A number of different materials have been used for this purpose. There have been numerical control systems built that use film, punched cards, magnetic tape, and punched tape as the data transfer medium.

Early numerical controls utilized punched cards because of the flexibility provided by their discrete nature.[2] They can be easily sorted, merged with other cards, or rearranged into many different sequences without having to be re-recorded. On the other hand, a card does not contain much information per unit area in comparison to various other media and is vulnerable to error in handling.

The high data storage densities and high read/write rates of magnetic tape prompted its use in numerical control. It is reusable and can be erased, recorded, and played back repeatedly without time loss. However, data volatility makes accidental erasure in the shop a problem, and data invisibility makes impossible visual inspection of the tape information.

The principal advantage of punched tape is that the means for handling the tape and for recording and reading of information are relatively inexpensive. The medium is continuous, so that recorded information is not restricted in its volume, yet a short message does not "waste" unused tape. Data can be visually read and do not require elaborate handling or storage means. Punched tape does, however, have less sortability than punched cards, and less data density and slower read/write rates than magnetic tape. By far the most popular medium today is punched tape, with magnetic tape being used in some specialized applications. Magnetic tape is also used in cases where extensive computer functions are performed. However, ordinarily, a magnetic-to-punched-tape conversion is made for input to the actual controls.

It must be noted at this point that the following discussion assumes the more classical numerical control example. There are instances where additional or different steps are taken because of special equipment considerations. For instance, computations or data conversions performed by a general-purpose computer or a specialized unit such as a director may intervene in the procedure.

Using Punched Tape. For either positioning or contouring, the machine instructions are most often put on punched tape

[2] William C. Leone, "Choosing Punched Tape Readers," *Automation*, April, 1966, pp. 99–104.

CODE	MISCELLANEOUS FUNCTIONS
m00	STOP
m08	COOLANT ON
m09	COOLANT OFF
m21	TOOL OFFSET
m23	MISC. AUX. 1
m24	MISC AUX. 2
m25	MISC AUX. 3 (Creep off)
m26	X, Y TABLE OFFSET
m30	END OF TAPE

CODE	PREPARATORY FUNCTIONS
g80	CANCEL CYCLE (Mill or tap off)
g81	RAPID
g82	DWELL
g84	TAP (Spindle reverse)
g86	MILL ON
g88	RETRACT TO LIMIT

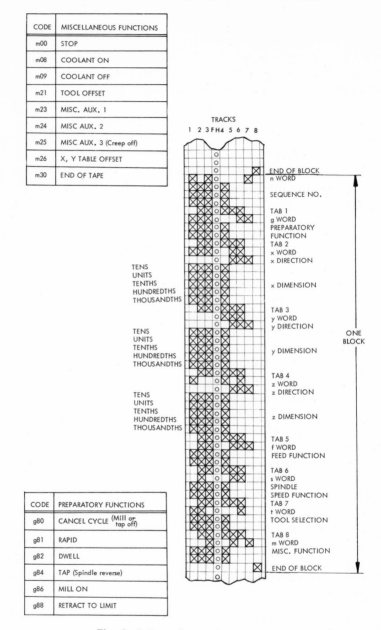

Fig. 2–6. Tape format for positioning.

for use by the machine system. The actual language, or definition as to what the holes in the tape represent, varies somewhat among the various systems. Input-output numbers or other symbols, or both, may be represented by different combinations of holes in various formats. However, significant progress has been made in standardizing at least some of the guidelines.

Standards have been proposed or established for the physical dimensions of tape used in numerical control and also for the language and format of information on tape. Proposed EIA standards [3] cover perforated tape with fully punched round holes used for recording six, seven, or eight levels of information across the tape. Dimensions and tolerances for the tape, holes, and locations of holes are established. In particular, format standards for numerical control systems for machine tools have been universally agreed upon.[4] The tape is 1 inch wide with eight information channels, and the formats for both positioning and contouring are set for most conditions. Generally, one character is represented by the presence or absence of holes in a line across the tape. The line is usually perpendicular to the normal axis of motion of the tape when it is transported. Some systems use a fixed number of these lines of information to represent a particular operation or other segment of data. The group of lines is known as a "block" of information.

Tape Format. Just as the planning sheet provides a systematic method of presenting the program in order to facilitate the preparation of the tape, so the tape format provides a compatible arrangement of the programming instructions for the control. It is the purpose of the tape format to systematize and give uniformity to the machining information so it can be properly utilized by the control.

Figure 2–6 shows a typical tape format for positioning control. Eight information tracks, numbered 1 to 8 from the edge of the tape nearest the feed (sprocket) hole track, are used to record the information. Tracks blocked in on each line of Fig. 2–6 indicate all the tracks that can be used on that line, depending upon the required information.

[3] EIA Standards Proposal Nos. 544, 571, and 588. Electronic Industries Association.

[4] EIA Standards Nos. RS-244, 273, and 274, Electronic Industries Association; AIA Standards NAS 943 and 955, Aerospace Industries Association.

Basically, the format requires that the entire program be broken down into individual steps, or operations. Each operation is described in one block of tape as in Fig. 2–6. So that the blocks can be easily identified, they are numbered consecutively by the sequence number that appears at the beginning of each block.

Comprising each block are specific groups of information called "words." Each word contains the instructions for a particular function. For instance, the positioning movement in the X axis is described by the X word, which contains the direction and dimensional information. In any given block, all words and all information in a given word must follow the format order, line by line.

One categorization of tape format is by the method of word identification. There are several, including "fixed sequential," "fixed block," "tab sequential," and "word address." The "fixed sequential" format assumes identification of a word by its location in the block. Words are in a specific order. In the "fixed block" format, the number of words appearing in successive blocks is constant. The two most often encountered are the "word address" and "tab sequential" formats. The "word address" format uses addressing of each word in a block by one or more characters which identify the meaning of the word. The type depicted in Fig. 2–6 is "tab sequential." Preceding each word is a tab, which is a line containing holes in tracks 2, 3, 4, 5, and 6. A word is identified by the number of tabs preceding it in a particular block. For example, the Z word is identified by the four tabs preceding it (its own and the g, X, and Y tabs). When no command is required for a given word, all the data for that word are eliminated, except for its tab.

To preclude any semantic confusion, the distinction between the first line in each word following the tab and the name of the word itself should be clarified. The information contained in line 1 carries no significance in the tab sequential format. However, owing to EIA standards, the line is included and reserved for those controls using other methods of identifying a word.

Numbers and instructions used with the various words and tracks they occupy are shown in Fig. 2–7. All numbers, including the sequence numbers, dimensions, rates, etc., are used in their binary form. This need not concern the planner (except

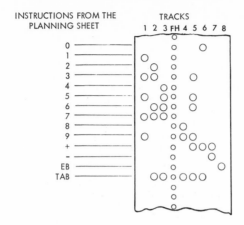

INSTRUCTIONS FROM THE   TRACKS
PLANNING SHEET   1 2 3 FH 4 5 6 7 8

Fig. 2–7. Tape-format details.

perhaps in checking a tape), since the binary arrangement of the holes is entered into the tape automatically.

Industry standards have established that each line of tape must contain an odd number of holes. When a line would otherwise contain an even number of holes, an extra hole is automatically punched in a special track (track 5) to bring the total holes punched in a line to an odd number. Thus, track 5 is called the "parity" track and is used for error checking.

The end of block command, the end of record command, and the sequence numbers are the only groups of information that do not depend upon a tab to identify them. The end of block command is placed on the line following the last line of programming information. If, for example, only a movement in the Y axis were required, the end of the block would follow the thousandths digit of the Y dimension. In this case, none of the remaining tabs in the block need be programmed.

Numerical controls use many systems for coding of dimensions. However, because punched tape yields its data by determining whether holes are present or absent, codes utilizing an on-off or two-valued, numbering system are required. Most systems, then, use a binary code, in particular a binary-decimal code, for dimensions.

Binary Numbering System.   A brief explanation of the binary numbering system is in order. In the following, it is given in the form of a comparison with the familiar decimal system.

Normal decimal counting begins with 0 and continues to 9. With the next count, two things happen: (1) the column containing the 9 (called the *units column*) goes back to 0, and (2) the next column to the left (called the *tens column*) becomes a 1, that is, 10. The right-hand column then starts over again and, with each succeeding count, goes one higher, until the 9 is again reached (19). It then goes back to a 0, while the left-hand column counts one higher (20). This procedure is continued—each time the units column goes from 9 to 0, a one is added to the tens column on its left. This continues progressively, adding a hundreds column, a thousands column, etc.

The same procedure is followed in the binary system, with one difference. Instead of ten digits (0–9), there are only two digits (0–1). Binary counting likewise begins with 0, but, instead of going to 9, it stops at 1. A 1, then, is the changeover point, just as the 9 is in the decimal system. With the second count, the 1 goes back to 0 and the left-hand column becomes a 1, giving 10. The next count makes the 0 become a 1 (11). With the fourth count, the two columns go to 0 and the third column from the right becomes a 1 (100). This double change may be compared with 99 going to 100. In both cases, the 9 and 1 have reached their limits in their respective systems and with the next count must return to 0. In the binary system, however, we have had only four counts, and, therefore, the binary 100 is equivalent to a decimal 4.

It should be noted that the digit in the extreme right-hand column is the least significant in both systems and increases in value to the left. Industrial standards for numerical controls, however, require that the least significant number on a punched tape be in the extreme left-hand column and that the numbers increase in value as the columns go to the right.

A second deviation is in the punching of the tape. Instead of taking the complete decimal number (such as 16.481) and converting it into a binary number, each digit is converted separately and placed on its own line. Therefore, it is only necessary to know the conversion of numbers 0–9. Figure 2–8 shows the binary equivalent of each decimal and the track on which it appears. A circle means a hole is punched. This method of listing decimal numbers is known as the "binary-decimal code." Because it is often expressed as "binary coded decimal," it is referred to in the industry as the "BCD system."

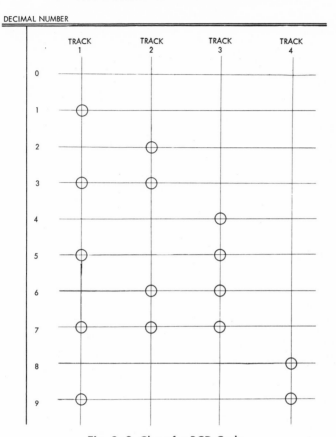

Fig. 2–8. Chart for BCD Code.

Preparing the Tape. Of course, the means of recording data on the data transfer medium is dependent upon the medium. Commercially available equipment that is "off line," that is, not integral with the machine and control system, is ordinarily used for preparation of the medium. For punched tape, the preparation unit essentially acts as a control for a tape punch. Several types of units are available, with keyboards ranging from fifteen push buttons to a typewriter style which types the program sheet and simultaneously punches the tape. The output of the tape-preparation unit is a tape containing the machining instructions to be entered into the numerical control system.

Checking the Tape. The possibility that incorrect information may be punched into the tape and thus initiate collisions, with concomitant damage to the machine tool, part, and personnel, is too great for most shops not to verify the correctness of the information. Only with relatively simple equipment or for very minor jobs is punched tape put to use without check. Then, the operator is apt to station himself close to the "Emergency Stop" button to shut down operations in the event he suspects a move is other than intended. However, regardless of the size and simplicity of the job, such a practice is dangerous and definitely not recommended.

There are a number of means for checking the tape, depending upon whether the type of error being checked is that of actual punching or that of programming. With regard to punching, there are several techniques available from computer art. Some are quite elaborate, tantamount to duplicating a large portion of the control unit for use in verification. Most systems assume use of the parity check mentioned in the discussion on tape format. It is a partial check only, but statistically more than adequate. Yet it is relatively inexpensive.

Another common technique is to inspect the tape visually for comparison with the planning manuscript. If the tape-preparation unit is of the typewriter variety, typed copy is produced corresponding to the punched tape. This is obviously easier to compare with the planning sheet, since interpretation of hole coding is not necessary.

Programming errors are even more serious and cannot be checked by equipment and procedures designed to verify only that the punching is correct. The easiest procedure, of course, is to make the part and inspect it for faithfulness to required specifications. However, making of the part is done one step at a time, which is known as operating in the "block mode." Correctness of moves is established by either actual measurement or comparison of display of position information by the control readout (if available) to the required moves.

Another popular technique is to make a trial run with a soft material or "air" substituted for the part. In this case, the machine makes all moves according to the tape data but the cutter does not encounter a part. Rather, collision programs result in either harmless damage to the soft material or "cutting of air." Again measurement or display means can be used for checking.

An *American Machinist* survey summary of plants using numerical control lists the primary method used by each plant (the total is greater than 100 per cent because of multiple answers):

Method	Per Cent Using
Tapewriter printout	34.9
Manual or visual check	13.9
Test run on machine	19.0
Dry run or substitute	19.5
Machine position readout	5.1
Inspection of first part	19.0
Other methods	7.7

Tape Preparation: An Illustrative Example

To demonstrate some of the foregoing, a fictitious sample part will be considered.[5] The part is shown in Fig. 2–9; assume it is to be fabricated with a turret drill and an incrementally programmed positioning control.

Before starting the program, it is necessary to complete the setup sheet. The initial step is to determine the drilling sequence that utilizes time most efficiently and still maintains the normal order of machining. The following spindles were selected:

Spindle	Tool
1	Center drill
2	¼″ drill
3	½″ counterbore
4	No. 7 drill
5	¼″-20 tap
6	½″ 4-flute end mill

Using this combination of spindles and drills, the operations will be performed in the following sequence: Center drill all eight holes; drill holes 8, 5, 4, and 1; counterbore holes 1, 4, 5, and 8; drill holes 7, 6, 3, and 2; tap holes 2, 3, 6, and 7; and mill slot.

This information is inserted in the setup sheet shown in Fig. 2–3. The remaining four columns, L, L_1, feed rate, and spindle speed, are determined from standard machine-shop practice. All numerical figures listed in these columns are average handbook values and are used only for illustrative purposes. It is expected that these figures would vary from planner to planner, depending upon the following conditions: the material to be

[5] Remex Electronics Numerical Positioning Controls, Programming Manual.

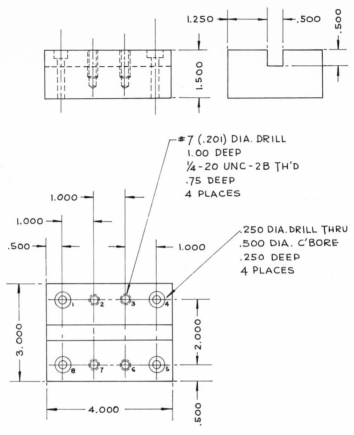

Fig. 2–9. Sample part for discussion of programming.

cut; the size, condition, and type of cutting tool; the depth of
the hole; the rigidity of the machine and holding fixture; the
selection of feed rates and spindle speeds available; etc.

Normally, it is not necessary to specify L, unless the depth
of the hole has to be held. In the example, holes 2, 3, 6, and 7
are held to 1.000 inch. Additionally, holes 1, 4, 5, and 8 are
counterbored to 0.250 inch. It should be noted that the L for
the counterboring tool includes an L_1, which is a 0.250-inch
pilot, making L for the actual cutting portion of the tool 6.750.

Programming of the tap cycle on spindle 5, in this example,
assumes the use of a positive-drive tap cycle with spindle reverse. Finally, the distance T must be given for one of the

spindles. This was done for the spindle with the longest tool (spindle 3) so as to facilitate the Z-axis calculations for the other spindles and to assure clearance during indexing.

After the constants have been specified, the program is determined and entered on the planning sheets. The first of the planning sheets is shown in Fig. 2–10. The program is normally started with the table at its right and forward zero limits and the drill in its upper limit. In particular, the first operation consists of positioning the table to bring the part under the tool and bringing the tool (spindle 1) down in rapid traverse to a clearance distance C. In this example, a clearance distance of 0.020 inch was arbitrarily selected; any distance may be used. The second operation brings the tool down to the work surface at a feed rate of 4 inches per minute (called out as code $f08$) and center drills a 0.250-inch hole ($Z = 0.020 + 0.250 = 0.270$).

A sample tape (Fig. 2–11) is shown to illustrate how the first two blocks of punched tape would appear.

EQUIPMENT

A numerically controlled machine system is made up of a machine tool and a control unit (Fig. 2–12).

Machine and Actuators

The machine tool is equipped to perform automatically by having drives and auxiliary function devices that respond to control by externally supplied signals. The mechanical means for automating a machine tool is a function of many variables. Consideration is given to the type of machine to be controlled, the degree of accuracy required, the versatility desired, and overall system compatibility. In general, the aim is to exploit the machine's capability as much as possible without undue complication.

There are three obvious methods for machine actuation, that is, for moving a slide of a machine tool:

1. Piston (direct coupling)
2. Rack and pinion
3. Lead screw

REMARKS	SEQUENCE NUMBER	PREPARATORY FUNCTION	DIMENSION +X	DIMENSION +Y	DIMENSION +Z	FEED RATE	SPINDLE SPEED	TOOL SELECTION	MISC. FUNCTION	END OF BLOCK
MOVE X AND Y INTO POSITION OVER HOLE 1, RAPID TO CLEAR	n001	g81	x+10500	y+06500	z+05970	f	s17	t1	m21	eb
	n001	g81	x+10500	y+06500	z+05980		s17	t1	m21	eb
CENTER DRILL HOLE 1	n002	g	x	y	z+00270	f08	s	t	m08	eb
	n002	g	x	y	z+00270	f08	s	t	m08	
QUICK RETRACT	n003	g81	x	y	z−00270 EB	f	s	t	m	
	n003	g81	x	y	z−00270		s	t	m	
POSITION X AXIS AND CENTER DRILL HOLE 2	n004	g	x+01000	y	z+00270 EB	f	s	t	m	
	n004	g	x+01000	y	z+00270		s	t	m	
QUICK RETRACT	n005	g81	x	y	z−00270 EB	f	s	t	m	
	n005	g81	x	y	z−00270		s	t	m	
POSITION X AXIS AND CENTER DRILL HOLE 3	n006	g	x+01000	y	z+00270 EB	f	s	t	m	
	n006	g	x+01000	y	z+00270		s	t	m	
QUICK RETRACT	n007	g81	x	y	z−00270 EB	f	s	t	m	
	n007	g81	x	y	z−00270		s	t	m	
POSITION X AXIS AND CENTER DRILL HOLE 4	n008	g	x+01000	y	z+00270 EB	f	s	t	m	
	n008	g	x+01000	y	z+00270		s	t	m	
QUICK RETRACT	n009	g81	x	y	z−00270 EB	f	s	t	m	
	n009	g81	x	y	z−00270		s	t	m	
POSITION Y AXIS AND CENTER DRILL HOLE 5	n010	g	x	y+02000	z+00270 EB	f	s	t	m	
	n010	g	x	y+02000	z+00270		s	t	m	
QUICK RETRACT	n011	g81	x	y	z−00270 EB	f	s	t	m	
	n011	g81	x	y	z−00270		s	t	m	
POSITION X AXIS AND CENTER DRILL HOLE 6	n012	g	x−01000	y	z+00270 EB	f	s	t	m	
	n012	g	x−01000	y	z+00270		s	t	m	

Fig. 2–10. Sample entries in planning sheet.

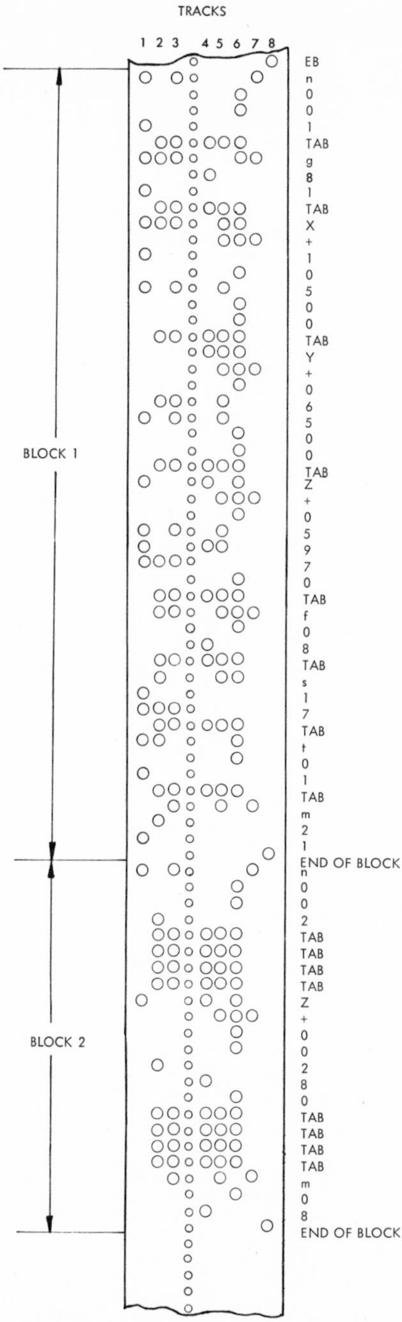

Fig. 2–11. Sample entries in tape.

45

The first method requires the application of a hydraulic drive; the others permit the use of either electrical or hydraulic control. Piston actuation has an inherent deficiency in achieving accuracy, because oil compliance makes it possible for no motion of the output to result until sufficient pressure is built up in the cylinder to overcome slide and seal friction. Since breakout friction may be much larger than dynamic friction, this effect presents a problem in overshoot if the desired position is only a few thousandths of an inch away from the current position.

Backlash in a direct-drive system is practically zero. Gear drives have inherent backlash, and elaborate efforts are sometimes required to minimize it. Preloaded split nuts are used on lead screws to eliminate backlash. The resulting friction is minimized by using ball-bearing nuts. Minimizing backlash in a rack and pinion drive is considerably more difficult.

Oil compliance presents a problem in direct drives (piston) with regard to ability to resist load disturbances (cutting forces). In general, compliance is minimized by gearing of the motor to the load.

Friction of the slide and hydraulic seals presents problems in any type of system. Assuming a gear reduction with 100 per cent efficiency, slide friction is reduced directly as the gear ratio.

For rapid traverse, the direct drive can exceed the gear drive in performance because its velocity is not limited by the gear ratio and maximum motor speed. A wide range of speeds is available with the direct drive.

These observations of qualitative comparisons among the methods of machine actuation are shown below:

	Piston	Rack and Pinion	Lead Screw
Accuracy	D	A	A
Backlash	A	D	D
Compliance (for cutting)	D	A	A
Ability to overcome friction	D	A	A
Ability to control inertia	D	A	A
Rapid traverse	A	D	D

A—Advantage
D—Disadvantage

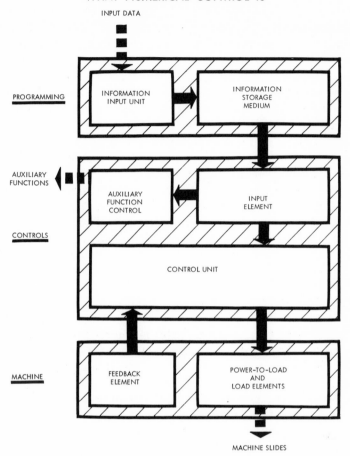

Fig. 2–12. Block diagram of machine system.

In summary, machine actuation by a lead screw offers several advantages over a piston or rack and pinion drive. This is borne out in the predominance of lead-screw-driven numerical control machines.

Drives for numerical control machines take many forms. The use of hydraulic, electric, or pneumatic drives is dictated by the type, requirements, and size of the machine tool and the type of control. The amount of servo control in existing installations runs the gamut from those with virtually none to some with very sophisticated regulation. There are simple machine-control arrangements, such as some which use stepping motors for the drives, where control signals direct the motors to step a prescribed number of steps corresponding to a desired displacement, and there is no feedback. A large number of positioning systems use a "bang-bang" or open-loop approach, where there is feedback of monitoring-device signals but the control is not a servomechanism in the most rigorous sense. In these, the drives are signaled to stop in accordance with control information based on feedback data from monitoring devices. However, servo action is of the on-off type, and, once the condition is established for signaling the drive motor to stop, further control ceases until the next operation.

Almost all contouring and several positioning systems use drive-monitoring-system arrangements that satisfy the definition of a servomechanism more accurately. The monitoring device continuously provides feedback to the motor in what is known as a "closed-loop servo" to result in continuous and proportional control of the drive.

Transducers

Attached to each machine slide is a position-measuring device for monitoring slide displacement or position and for feeding corresponding signals to the control. These signals are interspersed by the control in determining where the slide is or how much it has moved. The device attached to the machine slide is known as a "transducer." It is essentially a means for converting relative motion of mechanical members to measurable electrical (or other) signals directly proportional to the amount of displacement between the members.

Transducers for machine-tool control take on many forms, but all of them are coupled to the machine slides either directly or through gearing. In direct applications, the transducer is said to be "linear" and usually comprises a marked straight scale and a head for reading the markings mounted on the machine tool parallel to the axis of motion along which position measuring is to be done. The scale is fixed to the machine, and the head is fixed to the slide (or vice versa) so that relative motion produces quantitatively corresponding output signals.

Linear transducers have rotary counterparts whose physical workings are identical in every respect except that a marked circular disk plays the role of the marked straight scale. The package for housing disk and reading head and, therefore, the means of attachment also differ. Rotary transducers are usually coupled directly or through gearing to the lead screw used for driving a slide. In some rare cases, separate lead screws for transducer and drive are utilized if it is felt that coupling to the drive screw does not provide sufficient accuracy. In either case, the input shaft of the transducer turns in direct correspondence to the rotation of the lead screw, which is proportional to linear movement of the slide.

Rotary transducers are also coupled to the pinions in rack and pinion systems. Here, as in all applications of rotary transducers, rotation of the input shaft produces equivalent output signals. Consequently, the transducer is applied so that relative motion of the slide and the stationary machine member causes the transducer input shaft to rotate a proportional amount.

It was mentioned previously that systems may be absolute or incremental, depending upon how they are programmed. However, the absolute or incremental label is also a function of the method of feedback employed. "Feedback" is the name given to the transducer output because it represents signals "fed back" to the control in reporting slide movements. If the transducer gives output signals which can be interpreted merely as a numerical value corresponding to the amount of input shaft rotation, the feedback is said to be incremental. This exists most often with counting systems where a transducer shaft rotation produces output signals which are counted to indicate equivalent slide displacement. On the other hand, the transducer output may be such that, from a given starting point throughout its entire range, any input shaft condition is unique and different from any other condition. Such transducers are absolute in that the output indicates actual locations from a fixed reference as opposed to displacement.

Further categorization of transducers can be made according to the type of output signal produced. An analog signal is quantitative in nature and continuously follows the input variations. For example, turns of the shaft of a potentiometer result in directly proportional voltage levels such as are found in simple radio volume control. The quantitative value of a digital signal is nominally constant and is significant only in terms of the ability of the receiving or monitoring system to recognize

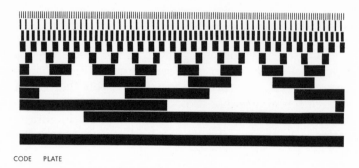

CODE   PLATE

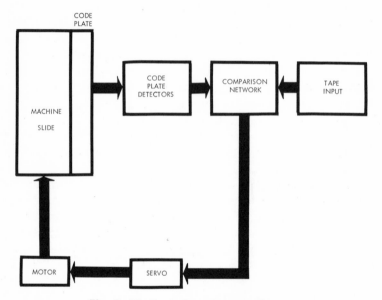

Fig. 2–13. Optical grating transducer.

whether it exists or not. Its form is discrete impulses, which are counted or otherwise interpreted to signify displacement or location.

Examples of transducers used in machine-tool control are many and varied. A few of the techniques employed are worthy of description merely to indicate the nature of typical implementations. It is not possible within the scope of this work to cover all of the variations in existence or, for that matter, to treat particular schemes unique to any one manufacturer.

SCALE

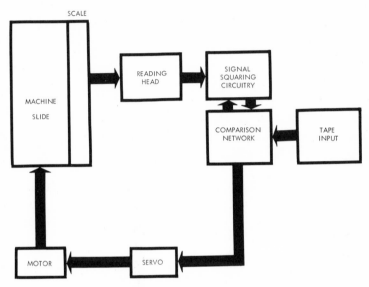

Fig. 2–14. Reflected-light transducer.

Direct Optical Reading with Gratings.   An early positioning system, such as is shown in Fig. 2–13, comprises a machine table carrying a glass strip that has 13 narrow tracks of alternately opaque and transparent segments. The segments are arranged in variable lengths and positions to conform to a code system. The simultaneous indications obtained from the thirteen tracks, at any position along the coded plate, never repeat the same combination. This happens to be known as a Gray code and falls into the category of "absolute" transducer discussed previously. The plate is read by 13 optical sensors' (phototransistors) looking through a transverse slit 0.001 inch in width. A lamp and simple condensing lens assembly illuminates the code plate opposite the reading slit. The reading from the coded plate changes for each 0.001 inch of machine table travel and always indicates the absolute position of the table.

A coding different from the Gray results if only one track is used in place of the several parallel tracks. Then a direct count of passing grating lines can be made, utilizing either transmitted or reflected light.

A reflected-light system (Fig. 2–14) may employ solid stainless-steel plate or tape carrying a pattern of lines running transversely across one surface. The lines may be etched into a highly polished surface to form a permanent pattern of alternate highly reflective and non-reflective lines. The lines are sensed and read photoelectrically through an analyzing grating in order to produce a fringe pattern. The output, when used with appropriate circuitry, is electrical square waves having a phase relationship dependent upon the displacement of the head along the grating. Phase comparison and counting techniques are used to provide the full range of measurement.

Optical Reading with Ronchi Lines. Two similar gratings made of transparent material may be superimposed on one another so that a small constant skew angle exists between the lines, thus producing a pattern of intersecting lines. The locus

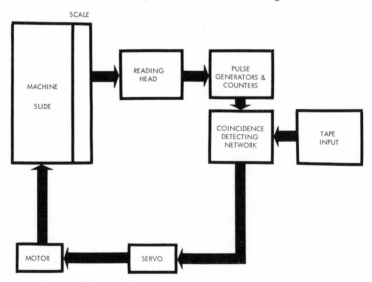

Fig. 2–15. Control for machine slide.

GRATING #1 (STATIONARY)   GRATING #2 (MOVABLE)

Fig. 2–16. Similar gratings with lines skewed.

of intersections appearing as a dark zone of low light transmission is called a "Ronchi line." Displacement between the gratings produces an apparent motion of the Ronchi lines proportional to the amplitude and related to the direction of the displacement. Optical sensors such as photodiodes can indicate the passage of a Ronchi line. Detection of both the presence and absence of light yields the maximum amount of position information from a given set of gratings.

One scheme uses a reflecting metal grating and a transparent grating with three apertures. In this arrangement, the amplitude of the displacement is indicated by a count of the number of Ronchi lines passing; the direction is indicated by the order in which the three apertures encounter a particular line. A similar scheme with transmitted instead of reflected light is also possible. Figure 2–15 shows one form of control system for a machine slide.

Figure 2–16 shows what happens when a grating is superimposed on another, similar one, with the lines of one grating slightly skewed from parallelism with the lines of the second

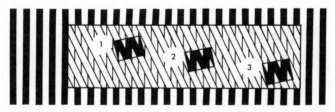

GRATING #2 (MOVABLE)   TRANSPARENT GRATING #1
                                 (STATIONARY)
                                 APERTURES AT 1, 2, AND 3

Fig. 2–17. Three reading apertures.

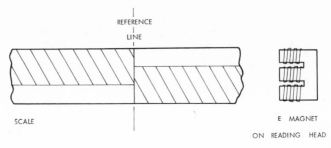

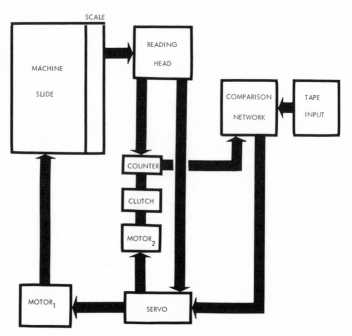

Fig. 2–18. Magnetic-element transducer.

grating. Figure 2–17 shows three reading apertures in an otherwise opaque grating (No. 1). If grating No. 2 is moved to the right, apertures 1, 2, and 3 will go dark in succession.

Locating a Large Step by Matching Magnetic Elements. In this arrangement, a magnetic element providing a sharp change in reluctance is attached to the moving table of a machine driven by a lead screw. The magnetic element is sensed by a magnetic pickup driven by a second lead screw. Because of the step, the magnetic pickup can detect if its location is immediately at the step or to a particular side of the step. A counter

Fig. 2–19. Inductosyn: *top*—rotary disks; *bottom*—linear scales. (Courtesy of Farrand Controls, Inc.)

on the lead screw driving the magnetic pickup yields its position. Commands from a tape input are used in combination with counter output to produce an error signal indicating departure of the table from the ordered position. A servo system positions the table so as to follow the ordered motion. Figure 2–18 is a diagram of a simplified system using this device.

Inductosyn (Farrand Optical). Silver conductors can be etched on glass disks to produce an accurate multipoled winding. The inductosyn is made up of two stators and one rotor of this type in a manner similar to a conventional synchro. By this means many electrical cycles are generated for each mechanical rotation of the shaft. Vernier capabilities with this method are such that high accuracies are possible with the rotary type inductosyn (Fig. 2–19 top). A linear version is also in use (Fig. 2–19 bottom).

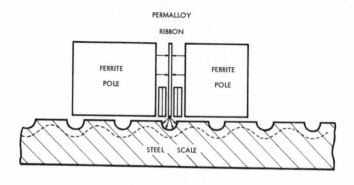

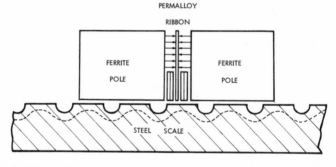

Fig. 2–20. Magnetic reluctance transducer. (Courtesy of Hughes Aircraft Co.)

Magnetic Detection on Steel Grating. A magnetic head can detect reluctance changes in a grating formed of etched or ruled lines on the surface of a magnetic material (Fig. 2–20). The grating used for this purpose is usually flat, but it can also be helical.

In its basic form, this method can be illustrated as a magnetic head near, or in contact with, a magnetic grating consisting of hills and valleys. The sudden reluctance change produces a corresponding inductance change in the head. The pulse so produced is then fed to an impedance-matching transformer and to appropriate amplifiers.

Vernier techniques using several heads can be applied with this method so that a displacement count density higher than the grating line density can be obtained. A flat grating is used in a manner similar to the corresponding direct-reading optical method.

Rotary configurations of the foregoing schemes are in prominent use because of the ease of mounting them directly to a lead screw (or precision rack) or coupled through a gear arrangement.

Locating the Transducer.   From an equipment-organization point of view, it is purely academic whether the transducers and other position-measuring apparatus, drive controls, and auxiliary function controls are considered part of the machine tool or the control unit. As more numerical control machines are produced, advance in design concepts are resulting in more truly integrated systems so that the lines of demarcation between machine tool and control are becoming less discrete. This is as it should be. In any event, the transducers, drives, and auxiliary function actuators are physically located on the machine tool, but the main elements of their control are usually in or near the control unit and are considered part of it.

Controls

From a control standpoint, the fully automatic operation of a numerical control machine-tool system comprises three basic phases—tape reading, controlling, and position measuring.

Tape reading signifies the conversion of the data represented by holes in the tape into information usable to the control. Tape readers are almost always integral to the control units. Their input is the tape containing machine instructions, and their output is signals to the controls.

The tape readers used for numerical control are most often "serial." A serial reader is one that translates the code combinations from a perforated tape by reading lines of data sequentially. Some "block" readers are used, that is, those which read a fixed number of lines at a time, but they are few in comparison to serial readers.

Tape readers may be mechanical, photoelectric, pneumatic, or dielectric, based on the method used for sensing the presence or absence of holes. Early numerical control systems used mechanical readers almost exclusively. The overwhelming trend has been to photoelectric readers because of their greater reliability and speed. Either they sense light reflected from the unperforated portion of a tape, or they sense light that passes through the perforations. The latter technique, which is the more prevalent in industry, has been refined to a very high degree of reliability. For all practical purposes, photoelectric readers have no speed limitations in reading. Because of handling and stopping requirements, the top reading speed is most often nominally 1000 characters per second. Since the holes along the axis of movement of the tape are on 0.1-inch centers, this speed represents tape movement of 100 inches per second. For numerical control, the reading speeds used are between about 10 and 500 characters per second. The trend is to increased speeds.

Position measuring, as mentioned before, is the monitoring of the positions of machine-tool elements at all times. The transducers for position measuring are attached to the machine members. However, the position-measuring circuitry for interpreting the feedback signals is usually part of the control.

The main function of the controls is concerned with information processing. There is always a portion associated with auxiliary functions where the control job is to receive signals from the reader, determine their proper sequence, and transmit commands to corresponding actuators. That part of controls normally called the "relay" (or "SCR") section is of this type. Their operations are often (but certainly not always) of the on-off type.

The more important information processing is associated with dimensional data related to slide movements. Controls may interpret output signals from the reader and simply direct drive elements to move accordingly. In most controls, displacements are achieved by commands given in association with comparisons made within the controls. The comparison may be between actual position and programmed position, count in a register and zero, count in a register and a predetermined number, or the like. In almost all of these, the information processing involves at least simple arithmetic adding and subtracting.

From an implementation point of view, controls may be electronic, hydraulic, or pneumatic. By far the most prevalent of the three is electronic control. From the earliest controls to current models, control design has followed the most advanced logic and circuitry techniques and, in fact, has led the way in certain areas. In general, the art follows closely the technology of computers. Early controls were of the vacuum-tube type. In about ten years, controls have gone through the conventional solid-state art and have been among the first complete systems to use integrated circuit techniques.

Numerical control systems have also been among the leaders in employing advanced packaging art. Modular construction, more commonly referred to as "building block" construction, has been used in numerical control since the early models. One reason for this is that a typical control can be divided into logical, separable modules very readily. For instance, one control has easily identifiable modules such as push-button and dial-in-panel, tape reader, position display, arithmetic unit, relay section, and power supply. This not only facilitates construction during assembly but makes for easier diagnostic techniques for maintenance and error checking.

CONTROL DESCRIPTION

In order to better appreciate how the foregoing control elements tie together and relate to the information flow diagram (Fig. 2–1) mentioned earlier in this chapter, it may be useful to look at a particular control in which these elements are readily identified. The one selected for this purpose is a numerical positioning control designed on the building-block principle with capability for controlling three axes simultaneously or in a specified sequence.

Figure 2–21 shows a control block diagram corresponding to the information flow diagram of Fig. 2–1. The numbers correspond to the same numbers in Fig. 2-22, where the physical

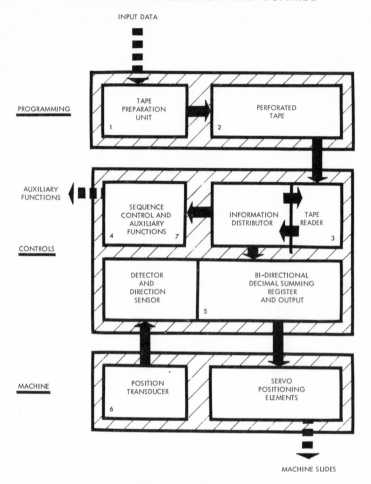

Fig. 2–21. Control block diagram.

elements themselves are pictorially represented. Finally, the control elements are shown in a cabinet in Fig. 2–23, where again the numbers identify the same elements.

With this control, the tape reader reads one complete operation—distance to be moved and direction of motion for each axis—into an element called a "bi-directional summing register." The output controls drive the axes simultaneously in the command direction at rapid traverse speed. The position transducers feed back digital signals indicating direction and distance of

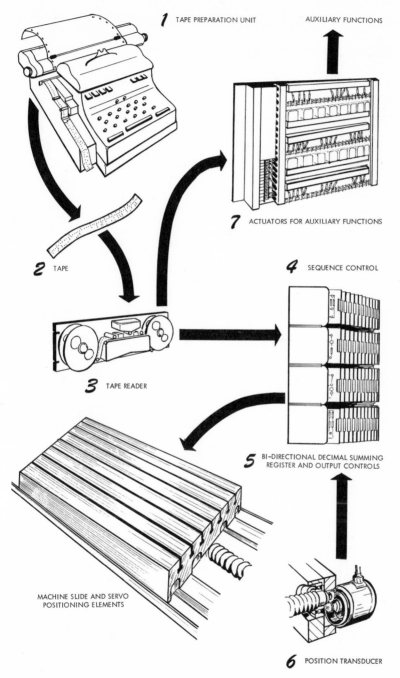

1 TAPE PREPARATION UNIT

AUXILIARY FUNCTIONS

7 ACTUATORS FOR AUXILIARY FUNCTIONS

2 TAPE

4 SEQUENCE CONTROL

3 TAPE READER

5 BI-DIRECTIONAL DECIMAL SUMMING REGISTER AND OUTPUT CONTROLS

MACHINE SLIDE AND SERVO POSITIONING ELEMENTS

6 POSITION TRANSDUCER

Fig. 2–22. Physical elements of controls.

Fig. 2–23. Control elements in cabinet: *left*—front view; *right*—rear view. (Courtesy of Remex Electronics.)

motion. The summing register counts the distance moved maintaining correct direction.

When the axis is close to the desired stopping point, the traverse speed is changed to a precise, controlled approach rate. When the axis is almost at the stopping point, the controls stop the axis motion.

The numbering of elements in what follows corresponds with that used in Figs. 2–21, 2–22, and 2–23.

1. *Tape Preparation Unit.* This is a keyboard device used in conjunction with a motorized tape punch to perforate tape.

2. *Tape.* Standard 1-inch-wide, 8-channel perforated tape is used.

3. *Tape Reader.* The input unit is a photoelectric tape reader operating at a speed of 300 lines per second so that essentially no machine time is lost for data input. Tape information is distributed to the bi-directional

summing register and to auxiliary function controls by electronic switching.

4. *Sequence Control.* The sequence control divides the system into two basic phases of operation—read phase and machine phase. The read phase is initiated by pressing a "read" button, which commands the input to read the first block of tape. When the end-of-block signal appears, the sequence control stops the tape reader and switches from read phase to machine phase. The start signal for machine phase arms the controlled approach and move actuators (relays), and slide movement ensues. When the slide has reached the programmed position determined by the bi-directional summing register and position transducer, the move actuator is de-energized and a coincidence signal appears in the sequence control. When all axes have reached coincidence, the sequence control switches from machine phase back to read phase and the cycle is repeated for every block of tape information. When the program is completed, an end-of-tape signal prevents the sequence control from entering the read phase again.

5. *Bi-directional Summing Register.* The summing register consists of counting decades to provide the resolution and travel required. Tape information is read into the register during read phase. This programmed dimension represents the incremental distance to the next slide position.

6. *Position Transducer.* In this case, the position transducer is a rotary photoelectric shaft-to-digital converter.

7. *Auxiliary Functions.* Functions such as spindle or tool selection, spindle speeds, coolant on-off, and rate of feed are controlled.

8. *Power Supplies.* The power supplies provide appropriate power for electronic circuitry and actuators such as relays. This control is similar to many in its amenability to operation in three modes:

Manual Mode—for operator control of the axes through dial inputs and jog controls.

Semiautomatic Mode—for one tape operation at a time under operator control. This mode is normally used for setup and for checking new tapes.

Automatic Mode—for completely automatic operation as described above.

3

Numerical Control Versus Conventional Methods

It is unfortunate that no manufactured part can be taken as being representative of all general manufacturing. Even in a particular industry, the variations in fabrication requirements militate against designating any one part as typical. Yet, just what numerical control can do in manufacturing is probably most effectively illustrated by comparative analysis of conventional versus numerical control techniques used on an actual part that has been manufactured by both methods.

There are many parts that fit this description. Numerically controlled equipment is new to many companies, so first applications are apt to be for making parts previously made by conventional techniques.

One such part is shown in Fig. 3–1. It was selected substantially at random from those in a plant that installed several numerical control systems at one time and, therefore, had to change over to numerical control production methods on many parts simultaneously. It is not intended in the following discussion to give every detail of the preparation for the manu-

Fig. 3–1. Part manufactured by conventional and numerical control machining.

facture of this part. Rather, the sequences of steps required for both conventional and numerical control production will be discussed. Of course, there can be considerable differences in procedure among various manufacturing plants. However, these are due mostly to differences in size and availability of facilities and quantity of personnel rather than in the fundamental operations performed.

Conventional Machining

The starting point in the manufacturing process is paperwork describing the work that must be done. This paperwork is prepared in a product engineering, or product design function, where the part requirement was first generated. A part drawing is made giving all of the vital statistics of the part, including material, dimensions, tolerances, finishes, etc. In a part such as the one under discussion, a drawing, or drawings, is also prepared for the rough casting, which must be made available to the shop for machining. Figures 3–2 and 3–3 show a portion of the detail on the part drawing. Besides the conventional three views usually shown for any part, a complex three-dimensional machining situation such as the one in this example requires certain section views to be shown also. Figure 3–4 shows one of the section views for the part.

Armed with the drawings for the part itself and for a rough casting of the part, a planning, programming, or processing function takes over. This function has the responsibility for

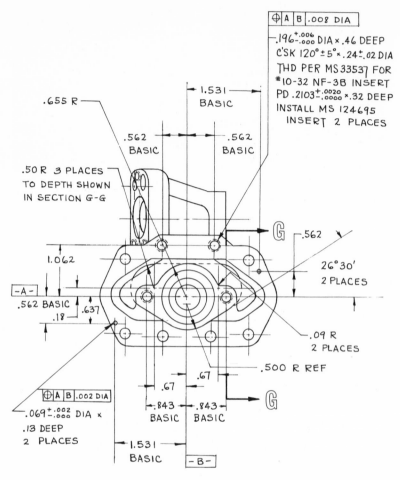

Fig. 3–2. Detail on part drawing.

making available to the shop all of the implements required for making the part. The rough-casting prints are sent to the appropriate department, or outside vendor, for making of the casting itself. The part drawings and a tool order are sent to the tool engineering and design function for making of the tools and fixtures required for holding the part during its machining. This involves the most important aspect of the planning function. The planner prepares planning or programming sheets, which describe the work to be done in the shop. These delineate

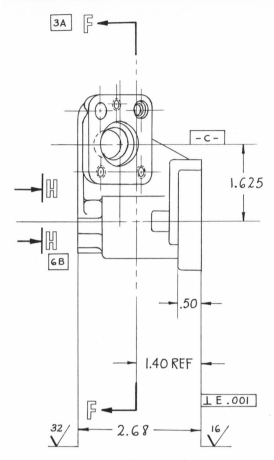

Fig. 3–3. Detail on part drawing.

the machining operations in some detail, usually arranged so that each sheet shows the operations that can be performed in a particular setup on a particular machine. Figures 3–5 and 3–6 show typical sheets from the set required for the part under discussion.

It will be seen that these planning instructions do not give the details of the tooling required. Rather, the tool engineer and designing function derive the information for designing the tools and fixtures from these sheets by determining how the part must be held to allow the operator to perform the machining functions delineated. Additionally, it can be seen

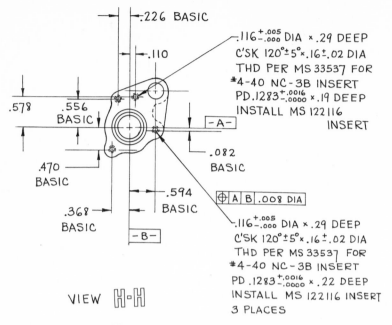

Fig. 3–4. Section view for part.

that the machining functions are described in terms of the final dimensions desired. It is left up to the operator to achieve these dimensions. For example, if a particular desired dimension is to result from cutting or milling 1 inch from a surface, it is left to the operator to decide whether he attempts to cut 1 inch from that surface in two 0.5-inch mill cuts, with five 0.2-inch mill cuts, or with some other combination, depending on his experience in using the machine and cutters and working with the particular materials at hand.

The tools and fixtures designed in the tool-engineering function are fabricated and sent on to the fabrication shop along with the part drawings, setup sheets, and planning sheets. In the shop, although the operator has all the implements required to finish the part and the instructions have been passed on to him, he still has many decisions to make. Basically, the instructions he has been given tell him how to locate the part in the tools and fixtures, how to locate the part on the machine, and what the final dimensions are to be. The setup sheets tell him what the tooling is to be. However, the cuts to be taken, the

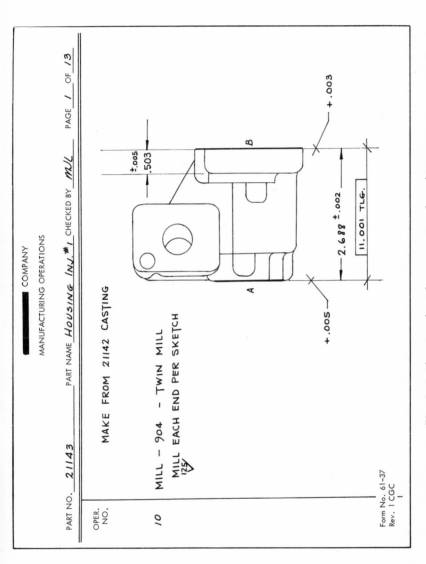

COMPANY

MANUFACTURING OPERATIONS

OPER. NO.	
	MAKE FROM 21142 CASTING
10	MILL – 904 – TWIN MILL
	MILL EACH END PER SKETCH

125

.503 ±.005

B

2.688 ±.002

11.001 TLG.

+.003

A

+.005

Form No. 61-37
Rev. 1 CGC

Fig. 3–5. Planning sheet for conventional machining.

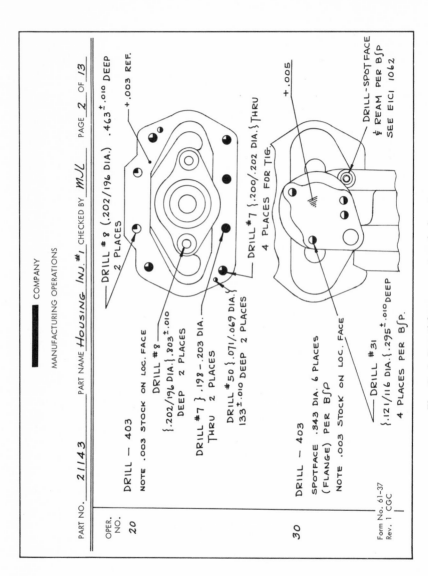

Fig. 3–6. Detail for conventional planning.

feeds and speeds, and the machine actuation are determined by him. Therefore, it is not unusual that a skilled operator will achieve a better finished product than will one who is not so skilled, although both may have been provided with the same tools and instructions.

Numerical Control Machining

Of course, in the numerical control machining situation, the same paperwork that was generated for conventional machining is usually available. However, since most shops with numerical control installations also have conventional machine facilities, planning often starts with decision making. A decision must be made as to whether the part is to go to the numerical control or the conventional machining facilities. Also, it is often the case, as with the example under discussion, that the numerical control equipment does not provide for complete machining of the part; that is, it may be necessary to have some of the operations performed on conventional machines. Therefore, the planning for numerical control machining will have interwoven among its instructions some for operations by conventional machining. This is, indeed, the case with this example. In the numerical control machining case, planning is necessary also to make available to the shop all of the implements required for making the part. However, the planning sheets take on a different character. Those that refer to operations by conventional machines will be similar to those aforementioned. Those that refer to operations by numerical control equipment are typified by the planning sheet one page of which is shown in Fig. 3–7.

It can be seen that the planning for numerical control gives more of the final statistics in the machining operation. The planner simulates the machining done in the shop. He may use a model or do it all on paper. He decides how the part is to be held and how it is to be located in the machine and determines each step required to perform machining that will result in the final dimensions desired. Accordingly, he goes through every step in very much detail, leaving no decision to be made later at the machine. The planner determines the feeds and speeds, the cutters required, and even the miscellaneous functions such as when coolant is to be on or off. Importantly, he determines

PROCESS PLANNING SHEET

Part Name: HOUSING INJ. #1   Date: Material: CASTING A356 ALUM.   Page: 1 of 17   Oper. No.: 1   Part No.: 21143

Seq. No.	Operation & Tool	Prep Comm	Longitude (X)	Vertical (Y)	Depth (Z)	Feed Rate	Table Index	Spindle Speed	Tool Code No.	Misc. Funct	Cont Feed Rate	Arc Center Offset X/X/Y	Arc Center Offset Y/Z/Z	Time
1	START POS. X & Y	00	101	05	0	8	0	0	0201	05	1			
2	TOOL CHANGE									06				
3	RESTART	00	101	05	0	8	0	23		03				
4	POS. Z				14445	8	0	23	0204	07				
5	MILL FACE A →		142			35								
6	RETRACT HOME				0	8								
7	INDEX POS. X & Y FOR FACE B		0879	05125			4							
8	POS. Z				14353									
9	ROUGH MILL FACE B →		1521			35		3						
10	POS. Z				14373			3						
11	FIN. MILL FACE B ←		0879			25								
12	RETRACT HOME				0	8		0		05				

Fig. 3-7. Typical numerical control planning sheet.

the feed rates and the depths of cut to be taken. Because the forte of numerical control is so often in their tool-change capability, the planner attempts to perform as many operations as he possibly can without change of setup. Therefore, he is largely responsible for the final design of the tools and fixtures, of which, of course, there will be less than in the conventional setup.

The particular sequences illustrated in the planning sheet of Figure 3–7 provide for positioning, rough and finish milling, and indexing of the part to mill surfaces marked A and B. Figure 3–5 shows the comparable conventional planning, which calls merely for the operator to twin-mill both surfaces at once, holding the overall dimension to 2.688 inches. Operations 4, 5, 6, and 7 are for milling one face (A), retracting the spindle, indexing the part, and positioning the X axis to mill the opposite side (B).

The completed planning sheets are then used for making the punched tape, which, along with the conventional planning operation sheets, setup information, and completed tools, goes to the shop for machining of the part. Figure 3–8 shows a section of punched tape corresponding to the operation and sequence numbers 4, 5, 6, and 7 of the planning sheet in Fig. 3–7. The dimensional information is represented in the standard binary coded decimal notation.

A Comparison of Conventional and Numerical Control Operations

For the part being discussed, three milling fixtures, three drill fixtures, and two lathe fixtures were used to perform the following conventional machining operations: five milling operations, two of which were profiling operations; twenty-five drilling operations including chamfering; fourteen tapping operations; six spot-facing operations; one precision-bored hole of multiple diameters (three diameters, three chamfers, two depths, and two formed undercuts); and three holes with step bores and undercuts. The same part was machined in a tool-changing numerical control system with one holding fixture to hold the part for milling, drilling, tapping, counterboring, and spot-facing, and a second holding fixture to hold left- and right-

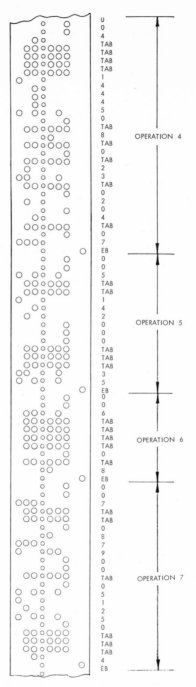

Fig. 3–8. Punched tape corresponding to planning-sheet entries.

hand parts for milling, drilling, tapping, and counterboring. With the first holding fixture, the operations performed were three milling operations, one of which was contouring; seventeen drilling operations including chamfer on each hole; eight tapping operations; six spot-facing operations; one precision-bored hole with multiple diameters; and one hole with step bore and undercut. The second holding fixture was used for operations on two parts, including two milling operations, eight drilling operations, six tapping operations, two large counterbores, and two small bores and counterbores.

Conventional tooling requires that the work-surface relationships be carried through from one to the other. For this reason, part tolerances have to be controlled closely to maintain locating references. Care must be taken in conventional tooling to ensure that the part is located securely in a way that will allow location to be repeated in case the same part has to be returned to the fixture. The design has to be such that locating of the part is foolproof.

Because of the limitations of conventional drilling machines, accuracy of hole location is dependent upon the drill fixture. If large counterbores are required in the part, it is necessary to provide slip removable bushings, which add about ¾ inch to the hole diameter. This sometimes restricts the drill fixture in the number of holes that can be put in an area. This may require that a second fixture be used to work the same area.

Conventional tooling usually involves more design and fabrication time than the tooling for numerical control because of the locating tolerance requirements. Engineering changes in the part may necessitate a new tool fixture in conventional machining to accept the change.

Since a family of tools for conventional machining is greater in number than that required for numerical control, the storage, maintenance, and necessary records required are a continuing expense. Setup of conventional machines is such that the job lot has to be a large one to offset the time involved. Scheduling of the job is more difficult because it must pass through so many departments before completion is possible. Additionally, conventional tooling and machine operation require many operator skills not always available in the shop, because of work loads and shortages of competent labor.

EVALUATING NUMERICAL CONTROL

Although many aspects of the foregoing example yielded conclusions typically reached in analyses for determining whether to use numerical control, not all potential advantages of numerical control are universally enjoyed in every application. In fact, in the foregoing example, the total machining times required for conventional versus numerical control methods were so nearly equal that they can hardly be used as the criteria for selecting numerical control. However, other advantages add up to rather spectacular efficiency and cost benefits. Before summarizing the advantages most commonly associated with the use of numerical control, it is useful to touch on several areas representing false notions concerning the criteria for evaluating numerical controls.

Production-Lot Size

Many are the times when prospective users summarily reject the idea of numerical control on the grounds that their production lots are not large enough. They hesitate because they feel that any type of automation is somehow of interest solely in mass production akin to the automotive industry of years past, where thousands of copies of the identical part were produced daily with no change for months on end. Their needs are for a small number of a greater variety of parts, they say, so they cannot justify numerical control. However, short-run production is precisely where numerical control contributes most. The facts that numerical control obviates the need for elaborate tools and fixtures and that changes are simply a matter of changing tapes make numerical control ideal for situations where the part varies frequently and the lead-time requirement per variation is short.

By the same token, it must be noted that numerical control is not restricted to the large manufacturers. The need for versatility in fabrication is universal. Many small shops have found that one numerically controlled tool takes the place of several conventional tools in terms of adaptability to varying part configurations, and, therefore, numerical control allows them to produce more efficiently and economically.

Tangible and Intangible Benefits

A common fallacy in evaluating numerical control is that the main reason for utilizing it is labor saving. This is, indeed, an advantage available with numerical control, but it is seldom the deciding factor in going to it. Labor saving accounts for a substantial decrease in overall production costs yet is only one of the many dividends users look for and get with numerical control. Tangible benefits over conventional machining include greater speed, accuracy, repeatability, ease and economy of process planning, savings in tooling, ease of setup, versatility of operation, economy in elimination of scrap, and shortness of lead times. In summary, numerical control's greatest assets are its flexibility and ability to effect dramatic increases in productivity.

Not to be ignored are the substantial intangible benefits from numerical control to which one cannot readily attach a dollar value. Yet, the indirectly resulting financial gains have been proved.

For example, analyses of parts to be programmed on numerical control tools often yield information about the parts useful in future design. On numerous occasions when part-evaluation surveys have been made to determine the appropriate type and size of numerically controlled machine tool to purchase, manufacturers have learned that certain steps can (and should) be taken to standardize machining sequences, hole sizes, etc. The result is not only economy in tooling inventories, but designs that are more readily amenable to fabrication by the machines in operation with little or no change of tool setup.

Information on the tape is complete regarding cutters, motions, speeds, etc., so times required to produce parts can be accurately determined. Consequently, production planning becomes more accurate and less costly, machine loadings can be made more precise, and unexpected idle machine time is eliminated. Production engineering, including forecasting of costs and production control is, therefore, facilitated.

Savings in tooling are straightforward enough. Just as important, yet difficult to assess, are the savings in inventory costs due to the lower required minimums in inventories because of the aforementioned shorter lead times between drawing and finished product. For example, the small shop with a requirement of 1200 of a certain part per year no longer needs to make all 1200 at one time to amortize tooling. Rather, 100 parts per month can be made simply by using the tape as convenient. This represents cash and space savings, which, in turn, represent tax and interest savings.

Having all instructions on tape avoids human operator error, since, once a tape has been checked out, it is correct every time it is used again for that part. In general, greater repeatability of the product results in greater part interchangeability, and improved quality results in inspection savings.

Computers

A commonly misunderstood point regarding numerical control has to do with the need for computers. It should be emphasized that the overwhelming majority of current numerically controlled machine tools are of the positioning (i.e., point-to-point) type. These do not *require* computers in any phase of their use. The drawings from which data are taken for tape preparation may be the same ones used in conventional machining; the planning sheet (also called the "programming" or "process" sheet) is prepared from, and contains information taken directly from, the drawing and requires only simple addition or subtraction plus standard data from charts on feeds, speeds, etc.; finally, the tape preparation consists of a simple planning sheet–to–keyboard operation on a unit much like a typewriter, which can be, and is ordinarily, operated by a clerk.

This does not mean that there is not a great deal of activity in the promotion of standard computer language in the tapes used with numerical control. The reason for this is that additional benefits are available with computer-implemented numerical control when certain repetitive subroutines are employed. However, these benefits do not require that the machine-tool user have a computer. Most of the recognized numerical control suppliers conform to tape-format standards that are quite generally accepted and are amenable to the computer programs if required. The important point here is that a computer is not at all necessary for achieving the advantages of numerical control and doing so now with available equipment.

ADVANTAGES OF NUMERICAL CONTROL

One of the best ways to understand what numerical control can do in manufacturing is to list the many advantages it makes available. Some of them are commonly mentioned in discussions on numerical control and are well known by the public in general. Other advantages are less well recognized.

Advantages available with numerical control can be put into five categories, which, for lack of better terminology, are listed here as productivity, flexibility, quality, administrative economy, and other intangibles. Those advantages accruing under productivity concern the reduction of labor or the time it takes to do a job. The flexibility of numerical control makes it possible to do a job with less direct equipment or to do more jobs with less equipment. Under quality we list those advantages that produce a better final result. Advantages in administrative economy provide for less investment for other than the direct equipment needs. Finally, there are other intangible advantages, which cannot be directly evaluated in terms of dollars or time.

In the following, no attempt is made to list the advantages in order of importance, since this varies with the application and the user's needs. Each item is shown under only one category, even if it is also applicable to others.

Representative applications of each of the listed advantages can be found in Chapter 7. In some cases, the benefits derived are attributable directly to the use of numerical control. There are others where the benefits come about as a consequence of improved machine tools whether numerical control is considered or not. However, the improvement features, in many instances, would not have been incorporated in the machine tools were it not for numerical control, which makes them economically attractive.

The long list of advantages possible with numerically controlled equipment should not lead to the conclusion that there are no pitfalls. Expensive and relatively complex equipment will have an audience ready to resist its introduction, so early missteps are costly in terms of complete acceptance. Additionally, certain organizational implications must be adequately recognized to ensure a smooth transition from conventonal to numerical control. However, proper planning for incorporation into company operations can make readily available all of the foregoing benefits and more.

Productivity

Automation is almost always associated with increased productivity. This category is used to itemize the benefits most commonly sought when automatic equipment is installed, that is, factors contributing to increased productivity through savings in labor and time.

Direct Labor. Savings in direct labor are almost always obtained with numerical control. In fact, this was one of the earliest reasons for its development. Unfortunately, this has been taken for granted to some degree so that many other advantages of numerical control have not been universally exploited or appreciated, owing to overemphasis, in some cases, on labor savings alone. Yet, labor savings can, indeed, be dramatic to the extent that economic justification for going to numerical control has often been adequately established on this merit alone.

Setup. Certain setup chores are eliminated or facilitated with numerical control. Some of the operations on tape preclude the necessity for functions which must be performed by the operator to set up the machine properly in the conventional situation. For instance, many conventional machines require that stops be set for certain critical dimensions. With numerically controlled machines, displacements are achieved from the tape information automatically with minimal setup effort.

Another advantage of numerical control in minimizing setup costs is the multioperation machining center, which can perform the work that conventionally takes several machines. Once the initial setup is made, all operations are performed without further need for loading, locating, and clamping, whereas, with conventional tools, some setup labor is expended every time the part is put on another machine.

Changeover. Numerically controlled equipment permits changes of jobs with minimum time lost between jobs. This is largely an additional dividend from the foregoing factor, simpler setup. This is also a factor contributing to greater flexibility with numerical control.

Machine Time. The automatic nature of numerical control machines allows higher speeds to be used. With conventional machines, accuracies depend on an operator's reactions in stopping a machine; to stay within allowable tolerances, operating speeds are often degraded.

Operator Capability. Numerically controlled machines are fundamentally easier to operate. The skills required of an operator are less than with conventional machines, where the operator must often be a trained machinist. With numerical control equipment, the operator must, of course, know his equipment. He must have the training and intelligence required to perform several rather straightforward prescribed routines, but he does not have to possess the technical skills of the experienced machinist. The intelligence corresponding to these latter skills is on the tape in numerical control. Of course, it may be argued that the skills are required after all, since the planner had to have this knowledge in order to prepare the tape. This is so. However, once the tape has been prepared, it may be used over and over without further involvement by anyone with these machinist's skills.

Decision Time. With numerical control, the manifold decisions represented by the data on the tape preclude the necessity for time to be taken on the floor in reading and interpreting drawings and in determining speeds, feeds, depths of cuts, etc.

Flexibility

As has been stated before, automation results in increased productivity. Very importantly, it also allows much greater flexibility of production. This comes about chiefly through the attainment of desired production levels with less direct equipment than would ordinarily be required.

Tooling. Savings in tooling costs are among the acknowledged advantages of numerical control. Automatically controlled machines do not require that drives and actuators be controlled by the operator. Therefore, handwheels, levers, etc., do not clutter the machines in the general workpiece area. In particular, most numerical control machines have index tables, which facilitate maximum accessibility of the cutter to the workpiece. This, in turn, facilitates the tooling compared to what is ordinarily required in the conventional case.

Location of the part and cutter in numerical control comes about as a consequence of tape-controlled machine actions. Location reference is determined during setup; thereafter, operations proceed from that point automatically, based on tape information. In conventional machining, accurate references must be established many times. The tools and fixtures for conventional machining are more complex and greater in number than those for numerical control.

Remote Tape Use. An increasing practice, made possible by numerical control, is to have similar items of equipment in any location operated by tapes prepared for any one of them. In fact, this represents one of the main targets of numerical control development, in which the military services have been particularly interested. This allows planning and tape preparation to be done in one place and the actual fabrication anywhere that the numerical control machines exist, with the bulk of the communications between facilities being the prepared tapes.

Prototypes. For almost any product, the making of the prototype is usually very expensive. Perhaps the biggest reason for this is that, in the making of a prototype, only one set of parts is made and advantage cannot be taken of the cost savings in quantity production. The beauty of numerical control is that small lots can be handled efficiently and economically. Thus, prototype parts are not prohibitive in cost. Furthermore, the tape used in making the prototype part can often be used, with whatever modifications are necessary, for the pilot or production models.

Mirror Images. Many numerical control systems are equipped to make either a left-hand or a right-hand part from one tape by simple selection of a mirror-image command. In conventional machining, the making of mirror images calls for complete planning and separate fabrication of each part. With numerical control, literally half of the planning is eliminated.

Quality

There are advantages in the use of numerical control that aid in producing a result of higher quality. These are far-reaching in importance because, in some cases, they make possible product features which cannot be achieved by any other means.

Accuracy. Certainly, there are many cases where accuracy is desired at almost any cost. Numerically controlled equipment makes it possible to attain extremely high accuracies in some cases where manual methods are restricted because of limitations in operator reaction speed and sensitivity to deviations from desired performance. The response of electronic controls is superior to that of humans, so that operation of machines within very close tolerances is accomplished more readily with numerical control in most cases.

It should be pointed out that there are instances in conventional fabrication where the iterative process is used in a long series of cuts by trial and error until a very accurate piece is obtained. However, this is contrary to what would be acceptable in production of more than one piece. The process is not only very tedious but is subject to the difficulties of having to check dimensions between trials. This involves relocating either the part or the cutter each time. If the work is to be done within very close tolerances, each relocation is an invitation to catastrophic error; that is, a cut may be made beyond the allowable tolerance, and the part then scrapped.

Repeatability. Although there are numerical control systems where one may question the absoluteness of the accuracies attainable, there is hardly any question regarding the repeatability usually obtained. The response of controlled machines to actuation signals is reproducible to a very high degree. By the same token, numerical control systems reproduce parts corresponding to desired specifications more reliably than is possible with manual methods.

This almost infallible reproducibility with numerical control systems provides another important advantage. Parts made by manual methods may all be within specified tolerances; however, there are, nonetheless, differences among them. In assemblies, tolerance buildup may preclude replacing one part with another. There are many cases involving assembly where a certain amount of matching is essential to assure that the parts will go together properly. With numerical control, faithful reproduction of parts allows parts interchangeability.

Administrative Economy

There are many costs in manufacturing besides those associated with direct equipment needs. Numerical controls are particularly useful in lowering required investment in these other areas.

Lead Time. In many industries, especially the automotive, a reduction of lead time from concept to finished product is very valuable in terms of marketing advantage. The shorter the lead time, the greater the opportunities to use the latest information on market needs and to incorporate appropriate features into the product before the design is finalized. With numerical control, the data on the tape indirectly are the product design. Consequently, to a certain extent, product-design changes are as simple as tape changes. The flexibility afforded by numerically controlled equipment in allowing product changes is reflected in greatly reduced lead times in comparison with those possible with conventional equipment.

Scrap. The yield in any manufacturing situation is the ratio of good products to the total quantity produced, and costs are inversely proportional to yield. With manual methods, deviations from specifications beyond the prescribed tolerances result in low yields. The aforementioned greater accuracy and reproducibility with numerical control inhibit scrap and rework so that yields are quite high. Operator-generated errors are practically nonexistent.

Inspection. The high degree of reproducibility with numerical control means that parts made from the same tape and on the same machine will be substantially exact copies. Therefore, once a part is qualified as meeting desired specifications, succeeding copies can be relied upon as faithful duplicates, except for changes in equipment or conditions. Many numerical control installations have made it possible for spot-checking to replace 100 per cent inspection.

Operating Conditions.   It was indicated before that numerical control allows cuts to be taken at higher speeds. In addition, positioning movements can be made at the highest speeds consistent with required accuracies. In fact, all operations can be performed under optimum conditions with numerical control. Because manual methods imply that each part may be made under different conditions dependent upon the operator, optimum operation probably occurs rarely. When a part is made by tape control, greater care can be taken in preparing the tape to ensure optimum conditions, since that same tape may be used repeatedly later.

Tool Life.   The optimum operating conditions possible with numerical control contribute to longer tool life. Perhaps just as important is the fact that tool life is much more predictable than in conventional manual operation.

Tool Inventory.   Since fewer tools are required with numerical control, there is a corresponding saving in the inventory costs associated with tools. Tools and fixtures are quite bulky and take up considerable space. Additionally, their usefulness depends upon their immediate availability. Therefore, storage for tools and fixtures is such that large spread-out facilities close to the shop are usually used, resulting in considerable cost. The numerical control counterpart consists of small, conveniently marked, and easily stored tapes.

Machine Utilization.   Because small lots are not incompatible with numerical control, machines themselves can be kept busy more easily. Idle time can be more readily avoided, since it is easier to fill with small-lot jobs or parts of large-lot jobs as required.

Space.   We have already touched on the space savings due to lesser requirements for tool inventory with numerical control. There are also space savings in the shop attributable to numerical control's smaller physical requirements compared to those of equivalent conventional machines. In addition to size economies resulting from the elimination of manual controls and accessories, further space is saved by the reduction or complete elimination of work stations required for the manual operators.

Investment Economy. Numerical control's lesser requirements compared to those of conventional machines apply to investment for the equipment a shop needs to do comparable jobs. Few conventional machines cost as much as their numerically controlled counterparts. However, in performance and capacity, the conventional machines do so much less that they are not true counterparts. The production from one numerical control machine sometimes equals that of several conventional machines. Consequently, for a given job, the total costs of numerical control equipment are ordinarily much less than the total costs of conventional equipment.

Product Inventory. Tool and fixture inventory has been mentioned. Much more important in most cases is the fact that product inventory can be minimal with numerical control. Shorter lead times and efficiency of small-lot production make it unnecessary to produce for inventory alone. Lesser requirements for inventory not only bring about direct savings in interest and taxes but also reduce potential losses due to obsolescence.

Handling Costs. Numerically controlled equipment performs many functions of transfer and indexing that normally require material handling with conventional equipment. Furthermore, since numerical control machines are equivalent to a greater number of conventional machines, handling between the machines themselves is also reduced.

Assembly. A large part of manufacturing cost is represented by cost of assembly. As mentioned before with regard to the merits of numerical control in reproducibility and accuracy, numerical control facilitates parts interchangeability. This, in turn, facilitates assembly, with a large return in cost savings.

Secondary Operations. Operating at optimum conditions makes it possible to obtain relatively good finishes so that secondary operations are rendered unnecessary or at least minimized.

Management Functions. An overwhelming advantage of numerical control is that so much of the production function is first done "on paper." This makes much of the performance predictable so that machine loading, scheduling, and costs are more precisely forecast than in the conventional case. Production planning is more accurate and less costly.

Other Intangibles

There are some advantages with numerical control that are not readily categorized but are valuable nevertheless.

Product Design.   Before making the decision to buy numerical control equipment, many plant managers analyze the parts they intend to produce with the equipment. Almost invariably, the analyses have led to similar conclusions regarding the product designs. When products are designed for conventional machines, hole sizes and other dimensions are selected arbitrarily to the extent that the part can function as desired. Analysis of parts for possible numerical control fabrication has yielded useful information on the product and the tools required to make it. In one plant, for example, analysis of a part showed that over 150 cutters were required to make the part. In determining how the part would be made by numerical control techniques, it was found that the part dimensions could substantially be standardized so that fewer than 30 cutters would produce the part.

Prestige.   One of the main requirements in proposals of any magnitude is inclusion of facilities and capabilities. For instance, most companies doing work with governmental facilities maintain elaborate accounts of such facilities so that the customer agencies have confidence in their ability to do work as required. In competitive bidding, the company with the most advanced facilities has an edge over those with inferior equipment at their disposal. Numerical control equipment has provided this edge to some companies in numerous cases.

The Factory of the Future.   Certainly, the computer is an essential ingredient of the factory of the future. This implies that many of the company functions encompassing not only all of the manufacturing departments but others as well will become interrelated via computer techniques. Vital to any such interrelationship is communications among the components of the whole system. If the machines in the factory are to be components of the overall system, they must be amenable to communications compatible with computer technology. The tapes of numerical control are, indeed, amenable to future linking with other computer functions.

4

Management
Implications

EFFECTS OF NUMERICAL CONTROL ON MANUFACTURING

In any discussion of criteria for evaluating whether or not to go to numerical control, consideration of effects of numerical control on manufacturing comes up. This involves some analysis of the various operating departments within the manufacturing organization. Unfortunately, there are too many varieties to permit discussion of the typical manufacturing organization. However, some general observations can be made.

Manufacturing Organization

Manufacturing comprises those functions directly concerned with yielding a product from an engineering drawing or description, plus ancillary, yet vital, functions including plant engineering and maintenance, facilities, personnel, manufacturing research, and other quasi-administrative activities. Our immediate interest is in the direct functions, particularly as they relate to parts fabrication. Furthermore, heat treating, painting, and other intermediate or finishing operations are not of direct interest in this discussion.

From an organizational point of view, a possible structure for accomplishing the manufacturing job may be illustrated as

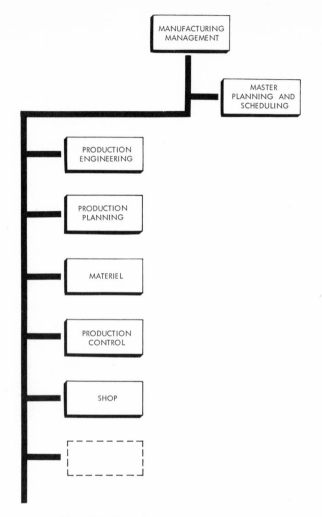

Fig. 4–1. Manufacturing organization.

in Fig. 4–1. Many plants are set up in just this manner. The intended functions of the depicted entities are sometimes combined in such a way that the responsibility hierarchy differs slightly from that shown. Of course, the organizational titles and lines and staff relationships may also differ. Nevertheless, the diagram is indicative of the main organizational elements encountered from engineering drawing to finished part.

Although there is some overlap in responsibilities assigned to the various organizational entities, nominal (somewhat simplified) areas of jurisdiction generally fit the following categories.

Production Engineering. This function is responsible for establishing producible part descriptions from the engineering descriptions. Very often, engineering drawings merely serve to describe desired parts without consideration for production equipment to be used, the process details, or even the limitations of available techniques. Careful analysis is necessary to make them amenable to fabrication utilizing equipment and facilities known to be available. Basic layouts are made so that the production process including tooling can be detailed. This may involve alteration of the details of a design (without change of function) and may include engineering and design of castings (if applicable) for the parts, material selection, and design of jigs, fixtures, gages, etc., that are special. The output of this department is production drawings.

Production Planning. This function is responsible for converting blueprints or other part descriptions into instructions for making the parts. This includes the specifying of machines to be used, setup data, and operation details. The refinement of operational steps for the actual production varies greatly among manufacturing plants. However, at least an outline of procedure is established for use by the shop. A very important function performed usually under the nominal jurisdiction of this department is the determination of tooling. In some cases, tooling is generated completely by this department.

Materiel. This function is responsible for making arrangements for providing the actual materials required to make the parts and storing them until required. This may include (along with others) make-or-buy decisions. Depending upon the in-house capabilities and work loads, the materiel function must constantly evaluate sources of materials required by the shop to implement production. Quite often a part is handled by the materiel department several times from its initial issue to its return into finished goods. For example, the materiel function may use drawings from the production engineering department to procure the raw casting from an outside vendor. The part may be alternately issued to the shop and returned to materiel jurisdiction (by being physically transferred from shop to stores, or, in some cases, even while the part is on the shop floor) after various operations are performed on it. Upon completion of all operations, the part may be moved to stores for ultimate shipment or transfer to the assembly function.

Accurate scheduling demands on the materiel function can be seen to require much more than initial availability of purchased raw material. In those cases where numbers of different parts undergo multimachine operations, smooth material flow is, indeed, a complex scheduling job. Although this is the responsibility of production control, materiel is intimately involved in the necessary communications and physical implementation.

Production Control.   This function is responsible for scheduling and coordinating shop load, that is, generally directing traffic of materials and instructions to the fabrication area. This requires a complete awareness of facilities, capabilities, and delivery requirements and involves management of the communications hierarchy among all of the contributing departments. In essence, it is the function of production control to optimize the match between demand (as represented by work load and schedules) and supply (as represented by materials, facilities, and personnel).

Shop.   This function is responsible for actually producing finished parts as output, using raw materials and instructions from other departments as input. This is the area in which facilities and personnel operate in accordance with instructions from production engineering, planning, and production control to implement fabrication.

Conventional Versus Numerical Control Machining

The purpose of the foregoing generalized chart is to facilitate the comparison between conventional and numerical control machining in manufacturing. In particular, there are functional counterparts of this organized setup for conventional and numerical control manufacturing that may be illustrated by Figs. 4–2 and 4–3, respectively.

Following the procedure from description to finished part for conventional machining, we see that initiating inputs to

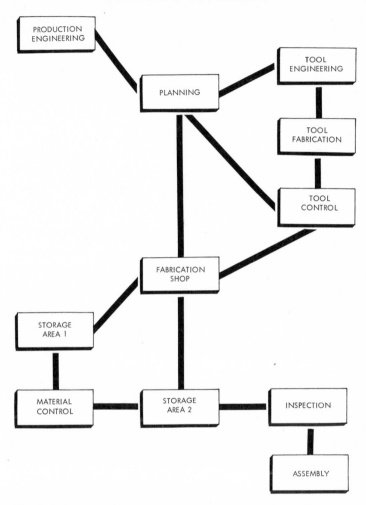

Fig. 4–2. Functional organization for conventional manufacturing.

the manufacturing organization are a manufacturing order and corresponding part descriptions, the latter originating from an engineering function. The production engineering department transmits its drawings to the production planning function and sends information relating to material requirements (such as for a casting, if applicable) to the materiel function, at all times communicating with the master planning and scheduling department. In materiel, a determination is made as to whether the material required is to be made or purchased. If it is to be

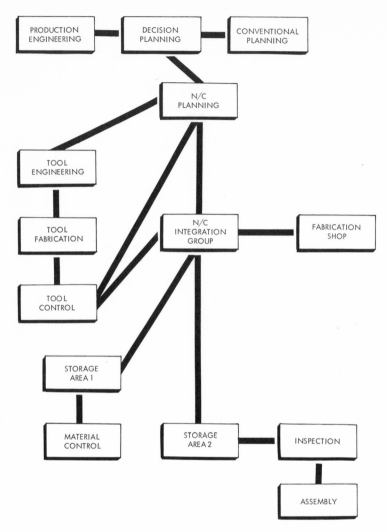

Fig. 4–3. Functional organization for numerical control manufacturing.

purchased, the order control function arranges for it. If it is to be made, material control may work back into the system for its fabrication requirements, in a manner similar to the scheme being pursued for the original part. The activity of either material control or order control is to have materials delivered to stores and made available for eventual use, as required, by production control. Meanwhile, the drawings are used in the planning function for determining machine setup, tooling requirements, and detailed operational planning. Tooling requirements and a corresponding order are submitted to the tooling engineering and design function, where appropriate drawings are made and sent to tool fabrication. Upon their completion, they are delivered to the tool crib and made available for use, as required, by production control. The planning function notifies production control when it has a complete package of setup, tooling, and planned part information. Production control analyzes this data along with established time standards and other information on traffic, machine capacity, available personnel, etc., to determine shop loading. As appropriate, the materials are scheduled into the shop from the tool crib and stores for parts fabrication.

It should be pointed out that the procedure outlined has been greatly simplified to apply most generally in parts fabrication. There are many feedback links among the functional blocks not specifically shown in the diagram of Fig. 4–2 in that many of the aforementioned steps are of a trial-and-error nature. Often, scheduling requires that the bulk of the outlined operations be carried out before castings are available. When castings are finally received, they have to be mated with the appropriate tools for proofing of the planning. Rework of casting, tooling, or detailed planning may be required. Sometimes, the machine initially planned for use is not available, so alterations must be made in the plan. Consequently, the diagram in actual practice may have many more connecting links representative of constant back-and-forth communications, depending upon the complexity of the part to be made.

Comment is also in order regarding the flow of drawings, instructions, data, and materials. Strictly speaking, the master planning and scheduling function and the production control function normally do not actually receive and transmit materials. In most cases, production control, for example, directs traffic, but materials, drawings, instructions, etc., change hands directly among the other functions.

Further comment is in order regarding the applicability of the foregoing to different-size organizations. In large companies, there may be entire departments comprising many people to perform each of the depicted functions. At the other end of the spectrum are the very small shops where all of the aforementioned functions are the responsibility of one man. In the latter cases, the communications among functions are sequential mental processes by the one man responsible, in accordance with a checklist perhaps.

If the connecting links between blocks are to serve merely to indicate the fundamental communications network among the main functions without regard to flow of materials, etc., the functional diagram for conventional manufacturing may be illustrated as in Fig. 4–4.

Comparison of Procedures. Analysis of numerical control manufacturing yields a very interesting result. Contrary to the widespread notion that numerical control inherently represents a departure from the traditional fundamentals of manufacturing, we note a striking similarity in that the diagram of Fig. 4–4 applies equally as well to numerical control. To be sure, there are differences; yet, the fundamental procedure seems to follow the same guidelines.

Where are the two not similar? Of course, we know the equipment is different. From a procedural point of view, this equipment difference is reflected most dramatically in the steps that make up the planning.

In the past, fabrication shops were generally staffed by machinists who were skilled in operations on all types of machine tools. The machinist classification implied proficiency and experience to the extent that a minimum of instruction was required in doing a job. Essentially, the machinist was given only the part drawing and told to make the part "according to the blueprint." When a job was released to him, he mentally planned how the job was to be done. Then he made, or had made, whatever tooling he felt he required. When his tooling was ready, he started the machining. Generally, he did all of the operations on each of the machines needed to do the job, regardless of type.

About the time of World War II, production requirements increased dramatically so that skilled machinists became a relatively scarce commodity. The lack of extensive apprenticeship programs in prior years in the United States aggravated the scarcity. At the same time, job lots increased to the point where machines were set up to produce the same part continuously.

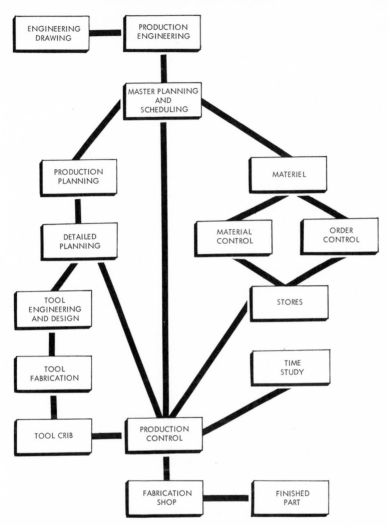

Fig. 4–4. Functional diagram for conventional or numerical control manufacturing.

In this way, machines could be operated by persons of more limited experience than the highly skilled machinists. Machinists were put to the task of breaking down the manufacturing process into separate operations to be handled by less skilled machine operators. This method provided a means for utilizing the machinists' knowledge and experience more widely. While the areas of manufacturing became more specialized, the need for coordination among areas became more important.

Planning, therefore, can be seen to vary from the completely informal mental process of the skilled machinist who is told to "make it according to the blueprint" to a highly formalized and detailed step-by-step listing of individual operations on each machine in every department. The conventional machining case may have either of these extremes of planning sophistication or any level between them, dictated largely by the size of shop, volume of activity, extent of repetitive operations, and degree of specialization. A small shop using conventional equipment is by practical necessity apt to do its planning less formally, that is, with less procedural documentation, particularly if all functions are performed by one person. As the size of shop and volume of activity become larger, there is a greater likelihood of specialization, and planning takes on more formalization until it approaches the level of rather complete documentation as with numerical control. Yet, even in a small shop, utilization of numerical control requires the formalized planning approach. It is not left up to the choice of the planner or operator. It must be done. The diagram of Fig. 4–5 schematically charts the relative degrees of formalization of planning for small versus large shops using conventional or numerical control techniques.

Care should be taken that this point is not misunderstood to mean that there is no planning in a one-man shop using conventional equipment, for example. In reality, every step must be predetermined, as it is in the large shop or for numerical control. However, there are two major differences. In the first place, the planning is not completely and systematically documented—it may be done entirely mentally. And very importantly, the planning may vary in details every time a job is done again by another individual.

It can also be seen from the foregoing that, in one respect at least, going to numerical control may represent a less dramatic change from traditional operations for the large shop. As mentioned before, considerations entirely independent of numerical control prompt the large shop to dissect the manufacturing process into a series of simple direct steps, which are then combined in sequential form on a document that represents the planning. Therefore, in the overall process, many large

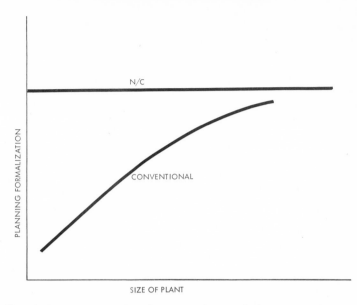

Fig. 4–5. Conventional and numerical control planning as functions of plant size.

shops have already done one part of the job even with conventional equipment that will be required of them with numerical control—the organized listing of operations. As systematic as this listing may be, it is still not as detailed as the programming required of numerical control, where every move (its magnitude, its direction, its duration, its rate, and how it is accomplished) is predetermined and set down in very precise terms.

Effects on Planning. The impact of numerical control on the manufacturing organization is perhaps felt most in a certain amount of shifting of the burden of getting the job done as efficiently as possible. The planner programs in much finer detail and takes on those duties previously assumed by the operator in determining even routine tasks at the machine. The operator's role with numerical control is sometimes relegated to simple loading and unloading of the part in the machine. The planner must be qualified in "seeing through" the complete job to finished part and cannot depend upon having an experienced operator intervening to ensure that the part is "made to print." The planner in numerical control, then, assumes a more important role. The very fact that he preordains every motion of the machine tool means he must possess full knowledge of part, fixture, and machine configurations. He cannot expect the operator to ensure that a cutter clears a protrusion on the fixture, for example. Therefore, although it is usually simpler than with traditional machines, the tooling is more completely determined by the planner with numerical control.

As mentioned before, step-by-step planning, which is vital to numerical control, initially came about largely as a result of the need for specialization in manufacturing even before numerical control equipment was generally available. This is almost paradoxical when one considers that numerical control equipment's versatility contributes to a tendency away from specialization in the shop. For example, it is not unusual in shops using conventional tools to find areas, departments, or personnel segregated according to type of machine—lathes, mills, drills, bores, etc. Numerical controls have made practical new machines, tooling, and operations for doing combinations of jobs without need for removing the parts from the machines. Therefore, more jobs are being done on one combination machine that in the conventional case required several unlike machines; thus, we see a trend away from specialization.

It is obvious from the foregoing that the trend away from specialization applies particularly to the planner, who cannot afford the luxury of having capabilities with only mills or only drills. He must be familiar with all categories of machining and, as a planner, must be a generalist.

The systematic planning and repetitive performance of numerical control are independent of the changeable characteristics of a human operator of the conventional tool. Consequently, labor inputs and, therefore, time standards are much more predictable. This affects the manufacturing organization with regard to where time study is done and how it is applied. Shop loading is easier to predetermine, which, in turn, means that production control is facilitated. In fact, the nature of data associated with numerical control machines is compatible with the form usually desired in computerized production control.

Because more of the tooling is specified by the planner, some of the decision-making responsibility of the tool engineering and design function is pre-empted. With regard to fixtures, numerical control requires simpler ones than conventional machines, or sometimes none at all. Since numerical control systems assume the job of part location that previously tooling had to assume, the main utility of fixtures is often merely to provide means for clamping the part to the machine table or carrier pallet.

Cutters, on the other hand, are usually more complex with numerical control. The reason for this is that the versatility of numerical control machines allows them to do more than one job, so that tooling people must make available combination types of cutters in many instances, including new and special designs of form tools, to fully exploit this versatility.

EFFECTS BEYOND DIRECT MANUFACTURING

The effects of numerical control on plant practices also appear in areas other than planning and tooling. In fact, they go beyond the confines of the manufacturing organization. Figure 4–6 shows the operating and staff functions, including those in manufacturing, where the implications of going to numerical control are felt, however slightly in some cases.

Plant Layout and Material Handling

A consideration of importance in evaluating numerical control is its effect on plant layout. Layout of both plant and equipment is a natural part of material handling. The advantages of single-story plant construction to facilitate moving of products was recognized as early as 1896. Soon thereafter, job shops began grouping machines together by type to improve production flow. In the boom year of 1928, the automobile industry invested $200 million in better plant layouts and modernized material handling. By 1935, three-dimensional layouts added new freedom in the analysis of material-handling problems.

Today, the determination of plant layout for smooth flow of materials is given very careful analysis, often involving quite sophisticated techniques. Proper placement of equipment must take into consideration that machines are to be adequately supported and reasonably convenient to power.

Both acquisition and maintenance of plant space are costly. Therefore, it is desirable to obtain whatever productive capacity is required in minimum space. Accordingly, machines must be located for maximum efficiency of material and work flow. Op-

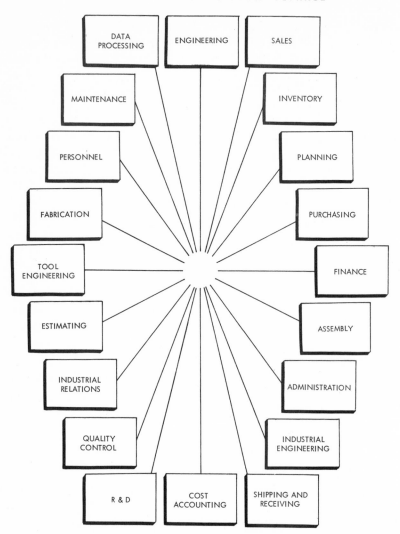

Fig. 4–6. Operating and staff functions affected by numerical control.

timum material handling implies shortest overall travel patterns with most convenient access at intermediate stations for auxiliary operations or handling.

The aforementioned sophisticated techniques have generated many improvements in plant layout and material handling in recent years. Yet, equally impressive progress toward these ends is being generated by numerical control. Numerically controlled machines do not have one-to-one correspondence with conventional machines, but it is generally true that a productive capacity equivalent to the latter's is obtained with a smaller number of numerical control machines. For example, in one automotive-components manufacturer's plant, five conventional machines—a milling machine, two drill presses, a boring mill, and a special-purpose machine—were used for milling, drilling, tapping, reaming, and boring. They were replaced by a numerically controlled combination machine—a machining center— with capabilities in excess of those represented collectively by the five conventional tools. The machining center occupies a nominal floor space of 21 feet by 9½ feet, or about 200 square feet, including the space in which the operator functions. The five conventional machines required nominally 250 square feet, with passageways between and around the machines, operator room, and materials-access needs increasing the requirement to an area closer to 450 square feet—more than twice that for the numerical control system.

Not all space savings are as dramatic as those in the foregoing example when numerical control machines replace conventional ones. However, reductions are generally large enough to be of substantial economic significance, especially when multiplied numbers of times in a large plant.

In addition to reducing floor-space requirements, compactness and multiplicity of operations performed per setup with numerical control machines help to reduce the actual cost of material handling. Consider the cycle from purchase and internal flow of materials through production control, warehousing, shipping, and distribution of a completed part. During only that portion of the time when the part is on the shop floor, it may be handled many times in being moved from machine to machine, in being loaded and unloaded, in being set up, in being oriented in various configurations for operations to be performed, and in being housed temporarily while it awaits its turn on a machine.

Numerical control eliminates a lot of handling, since multiple types of operations on one machine preclude the need for going from one machine to another of the single-function type.

Material handling may, and usually does, represent an important fraction of the product cost. For example, a hydraulic-component manufacturer who carefully analyzed the material-handling cost for several nominally typical parts in his production found it to vary approximately from 6 per cent to 20 per cent of the product cost, depending upon the particular item. Further analysis of total numbers of produced items revealed an average of 16 per cent of product cost devoted to material handling. When this manufacturer went through the justification steps for his numerical control machines, he gave specific attention to the possible handling savings. He found that, on these "typical" parts, handling savings attributable almost solely to the numerical control machines would be close to 25 per cent, yielding an average 4 per cent reduction in product cost!

Materiel

In conventional machining, it is usually desired that production lots be as large as possible to minimize inordinate time loss for setups. The result is stacking of parts at each machine until they can be worked on by the next machine in the production sequence. Not only are there costs associated with material handling between machines, but there are also costs in inventorying these parts on the floor at each machine. In either case, containers are required for storing parts as they come off a machine and for transporting them to the next operation. Containers used in conjunction with numerical control are sometimes of special design, but there are less of them.

For the reasons mentioned in the preceding paragraph, the monitoring of inventory takes on a different aspect. Whenever a physical inventory is taken in a plant, the work in process represents the most elusive type of inventory to which a dollar value must be assigned. Certainly the values of raw materials and finished goods are better-known quantities than work in process, which can be at many different levels of unfinished state. Since there are fewer parts at the intermediate levels of completion with numerical control, not only is there less inventory overall, but the dollar value of that which exists is easier to determine. In general, the shorter lead times from raw material to finished product brought about with numerical control substantially reduce the "tally clerk" types of responsibility in the materiel function.

The shorter lead times made possible by numerical control often enter into the make-or-buy decisions. It is standard policy in many plants that as many needed items as possible be made in-house if it is economical to do so by contributing to the spread of the expenses of overhead. However, short delivery requirements may preclude doing the work in-house with conventional tools. Therefore, the purchasing function has additional potential sources of procurement opened to it when numerical control is available.

Inspection

For particularly close-tolerance work, assembly may necessitate matching of parts. In a large assembly line, this can involve an intermediate step with personnel assigned to do this matching and its attendant record keeping. Numerical control virtually eliminates the need for this type of work, because of the precise repeatability and the resulting interchangeability. Obviously, the job of inspection is also easier, since parts are more consistent. First parts are checked thoroughly for accuracy of manufacturing and then inspected only on a spot-check basis. The quality-control function changes its emphasis from the traditional routine. Once the first acceptable part has been obtained, normal operator intervention cannot generate a deviation from the specifications. Factors such as tool wear can cause variance, but a very large source of variable error has been eliminated. Therefore, inspection routine can be more demanding on the first part and less so on succeeding parts. This seems to suggest that "pilot error" is eliminated with numerical control. With respect to the operator, this is true in most cases, which means the major source of variable error is indeed avoided. However, "pilot error" of a different type is too often encountered—programming error. On the plus side is the fact that programming error is not variable, so that it is more easily discernible and, once corrected, does not recur. This is the reason for the first part's being given such a thorough inspection.

The possibility of programming error does imply that personnel on the shop floor must become conversant with tape language in order to challenge data input to the system. This familiarity can be a required qualification of the inspection personnel, the operator, or both. Unfortunately, lack of simple interpreting ability on the part of shop-floor people has been costly in many plants where numerical control was introduced without complete understanding of its merits and pitfalls. For some parts where lot runs have not been particularly large, some users have not insisted on a first-part tryout involving as much care as is usually desirable. Tapes produced in planning were transmitted to the shop for production use with bad parts often the result. Deviation from specification often led to the conclusion that equipment had failed. Consequently, maintenance personnel were often called in at considerable expense to the machine manufacturer, the control manufacturer, and/or the user, only to find an error in programming. In such cases (which, incidentally, still occur too frequently), the situation is almost always aggravated by the fact that inexperienced floor personnel are too quick to attempt controls adjustments before the cause of the difficulty is ascertained. Therefore, by the time maintenance personnel arrive on the scene, machine and control conditions that existed when the difficulty was encountered are no longer present. In fact, the aforementioned attempts to make corrective adjustments may well introduce new problems so that maintenance people are confronted with a difficult diagnosis.

Maintenance

Whether maintenance for numerical control systems is carried out by in-house personnel or not, the routine differs from that for conventional machines. In the first place, new skills are required. The nature of controls is most frequently electronic, so some familiarity with this technology is desirable. Procedures have been developed that preclude the need for electronic engineers to do maintenance work. Most controls manufacturers provide operations training and manuals to help the user recognize symptoms of possible difficulty. Routines have been established for straightforward checks corresponding to each symptom so that fault diagnosis is a systematic procedure. Check tapes, which contain data specifically designed to highlight non-optimum controls adjustments or conditions, are often available. The modular characteristic of most controls facilitates fault correction. In fact, for all but the most serious controls problems, correction is even simpler than determination of the source of difficulty.

Preventive maintenance has become a very useful concept for numerical control users. Its importance is due to the potential catastrophic result of a controls failure that might cause a collision between massive, high-powered, and high-speed machine elements and cutters. Electronic controls are made up of literally thousands of components, so occasional failure of one is not unusual statistically. Of course, not all are critical, and failures of some can cause little more than inconvenience or perhaps some operational delay pending their replacement. In any event, the high cost of numerical control equipment makes it paramount that it be in use as constantly as possible for maximum return on investment. Therefore, prevention of machine and part damage, and avoidance of long interruptions of operations for fault finding and correction, is of utmost concern to the user.

Fortunately, the controls of numerical control systems are amenable to periodic checks designed to reveal marginal components. Therefore, systematic programs for going over the equipment at regular intervals to check critical subassemblies and components and do routine cleaning of components are becoming standard practice with numerical control. Foreign material in a shop atmosphere can sometimes interfere with optimum performance of the equipment, so the cleaning operation is important. The advantage of such a systematic program is that preventive maintenance checks can be of short duration and can be scheduled to minimize loss of production on-time.

A service arrangement for the controls is offered on most numerical control systems by the machine-tool manufacturer and/or the control builder either directly or through the vendor supplying the machine. Based on the aforementioned premise that periodic maintenance will add materially to the life of the control and will assure its proper function, the typical arrangement is designed to provide for periodic inspection and service of the controls in the plant of the user. The service furnished

under the terms of the arrangement usually includes all necessary adjustments, cleanup, and lubrication to maintain a good
operating condition.

The vendor in this case provides a specific number of factory-
service-engineer calls per year at the user's plant at predetermined intervals, for instance, six calls per year at intervals of
approximately two months. On each call, the service engineer
usually

1. Checks the position-measuring system
2. Cleans and adjusts the tape reader
3. Checks and records all power-supply voltages
4. Checks positioning of machine elements such as the table,
   spindle, etc.
5. Checks all functions of the control by tape input (with a test
   tape)
6. Changes air filters and gives the unit a general cleaning
7. Checks all functions of the unit manually
8. Checks the transducer and lubricates gearing (if applicable)
9. Inserts into unit and checks spare modules that may have
   been repaired since the last maintenance call

When the service engineer has completed his call, he prepares a report listing the work done and any replacement components used. This is necessary for accounting purposes but
also serves as a record for later reference as to machine performance.

Usually, the arrangement also includes provisions for emergency service, that is, service other than scheduled preventive
service. This procedure is usually put into a formal agreement
listing the specific equipment covered and the rates to be
charged, which, of course, vary with the number of machines,
their locations, and the frequency of calls.

The foregoing happens to deal with the main elements of
an arrangement between a controls manufacturer and a user.
However, the program applies equally well if the procedure
is followed by in-plant maintenance personnel.

A by-product advantage of preventive maintenance is that
machine operators are aware of scheduled checks, so they are
less inclined to call for emergency help when it is not absolutely
needed. If a serviceman is due next week to do preventive
maintenance, minor deviation from optimum operation today is

less apt to prompt a call for immediate help. Furthermore, regular expected visits from a serviceman, in-house or external, provide the means for useful exchange and discussion of observations and questions regarding operational characteristics of the system.

Organizationally, the maintenance requirements of numerical control point to the evolution of a new type of serviceman. The expert mechanic who can tear a machine tool apart and put it together again is not adequate if he cannot interpret simple signal patterns on an oscilloscope. On the other hand, an electronics expert without an understanding of straightforward machine concepts such as backlash and lead-screw windup will not do either. A numerical control system is an integrated package whose performance is a consequence of the machine tool, the controls, and their interconnections. Yet, deviation from optimum performance, which may result from a problem within the system, does not always identify the problem as mechanical or electronic. Therefore, familiarity with the technology involved in either the machine tool or the controls alone is not sufficient. The numerical control serviceman does not have to be an expert in either field, but he must have at least an elementary background in both.

Personnel

Whenever new skills are required, the personnel function is always concerned in the hiring, laying off, training, and retraining of people to do the work. One area receiving much current attention is that of personnel adjustment to some of the problems introduced by numerical control. Just as with any other technological innovation, there is some worker apprehension that numerical control will create unemployment. Just as important is the fear by the worker that, even if his job security is not threatened, his role will be that of some sort of robot with no opportunity for individual expression and identification with the final results of his labors—another example of the so-called trivialization of work [1] whereby man is relegated to the role of a mechanism of the most elementary type.

[1] Adam Abruzzi, *Work, Workers and Work Measurement* (New York: Columbia University Press, 1956).

It is true that, like other technological innovations, numerical control prompts the need for some adjustments. Unfortunately, the work force most affected immediately is that with the lower skills, and not all of the personnel involved can qualify for possible upgrading. In some cases, the retraining required would be of too long duration to be satisfying to either the employee or the employer. Therefore, some employees are given opportunities for other work when numerical control eliminates their specific jobs. The changes available may, indeed, not reflect upgrading. Thus, we must acknowledge transitory dislocations some of which do not represent job amelioration to every individual. In the overall, however, numerical control has proved time and again that new opportunities for upgrading and promotion of present staff and greater job security for the majority of the personnel, new and old, are opened.

Industrial Relations

One of the functional areas drastically affected by the introduction of numerical control is that of industrial relations. The reason for this is that the aforementioned new skills cross traditional classifications, and, in some cases, replace them. One operator properly trained can run or monitor two or more numerical control machines if the arrangement is compatible. The first reaction is that less work force is necessary, so labor repercussions are possible. Some shops using numerical control have followed conventional patterns because they have not resolved the relationships with labor and, consequently, they have not realized maximum savings. Actually, in most cases, the total work force is not decreased. The number of workers directly involved with production may be reduced, but often corresponding additional personnel are required elsewhere. Nevertheless, there can be dislocations during the transition to numerical control, which, in turn, can disrupt labor relations.

A pertinent example of the transitional problems that can be created occurred at a division of a West Coast aerospace company. The particular division under discussion was enjoying good business, so it was in the process of rapid expansion of facilities and work force. Management decided to acquire the organization's first numerical control machine, which, upon delivery, was to be installed in the plant among existing conventional tools. Space in the plant was reserved, and preparations were initiated for equipment installation. Of course, such preparations were obvious to the regular employees, who naturally speculated as to the equipment to be installed and its possible impact on them. Because the division did not have people in its work force with ready experience for numerical control work, and to preclude further strains on an already overloaded schedule, it was decided not to divert workers from the conventional machines. New people were hired for the prospective equipment. Members of the existing staff became apprehensive and perhaps logically so. After all, they had a quite common erroneous impression that numerical control is installed chiefly to reduce labor; they were not given an opportunity to participate, but, rather, new people were hired; the hired employees had new classifications with higher rates for jobs seemingly equivalent to theirs on conventional tools; and, finally, in management's well-intended efforts not to upset production schedules, information on initial preparations for numerical control was not fully disseminated; that is, some work planned for numerical control was actually segregated from other work. Of course, the regular work force thought and expected the worst. After the numerical control system was delivered, the period of installation, indoctrination, and tryout turned out to be unduly long. An inadequate foundation for the machine tool that contributed to a prohibitive sag of one end of the machine base generated ridicule of the new system among the regular work force before power was even turned on. During checkout of the system, several "pilot errors" resulted in many reject parts, which, stacked in a pile next to the machine, stood as a farcical monument to what seemed to be a bad management decision. The grand finale to this comedy of errors was a thundering collision of the machine spindle into the worktable. Eventually, the system was "debugged" and the numerical control staff worked smoothly as a team. Good parts became routine, and there was a growing realization of numerical control's superior productivity. During the several months that all of this was taking place, the company lost a few contracts and discovered it had overhired in an overoptimistic anticipation of sales activity. Consequently, for reasons entirely separate from, and independent of, the numerical control interest, the company was compelled to call a layoff. The man on the floor was convinced the layoff came as a result of numerical control. Naturally, labor problems ensued.

Perhaps this case represents an extreme set of circumstances, but there have been similar occurrences, at least in part, at other companies. The relationship between classifications of people working on numerical control equipment and those on conventional machines must be resolved. The opportunity for upgrading from conventional to numerical control must exist.

One aspect of industrial labor relations that becomes a union concern is that some categories of personnel may change from hourly to salaried classifications. In particular, this has happened with respect to planners and programmers in some plants.

Another important consideration is that, with traditional machines, operators are quite often categorized (and upgraded sometimes) according to the machine type. Some plants follow the practice of starting out new and inexperienced operators on simple drill presses and progressively advancing them to mills, lathes, grinders, and similar machine tools. With numerical control, these machine distinctions are not well defined, and the shop talent must be more generalized.

Industrial Engineering

Since numerical control implies greater conformity of shop activity to predetermined time-oriented programs, the time-motion study function of industrial engineering becomes less important. However, there is greater responsibility in the contribution the industrial engineering function must make toward recommending and justifying new equipment. Equipment is purchased not only to meet new requirements but also to replace older machines. Therefore, there are many instances of numerical control machines's being recommended to replace conventional ones. Justifications for the two kinds of purchases are based on different guidelines, so industrial engineers must familiarize themselves with the new concepts.

Additionally, the higher cost of numerical control systems, which is representative of their doing more of the total job, implies higher overhead. Frequently, the prospective investment commitment is of such magnitude that higher levels of authorization are required than is normally the case with conventional equipment. Consequently, initial and continuing costs of such machines prompt the corporate management function to become even more greatly involved in capital-equipment evaluations.

Product Engineering

One of the most important effects of numerical control on engineering is its direct influence in standardizing practices. Numerical control machines are more versatile and more general in their scope of capability. Therefore, the engineer has a better appreciation of the equipment available to fabricate the parts he designs. This consciousness is often reflected in designs that can be essentially completely produced on a minimum number of machines. Thus, the desirability of minimizing the variety of cutters required leads to engineering attempts to standardize on design parameters such as hole sizes.

Another area profoundly affected by numerical control is what has been variously called "production engineering" or "process engineering." In the past, the product and manufacturing functions have been rather completely segregated, so that the machines and procedures for production often have been organized without adequate early planning. Automatic production, especially as represented by numerical control, makes available dramatic economies and efficiencies only if it is properly evaluated, planned, and engineered. Consequently, a new function has evolved and has been given the name "manufacturing engineering." As Roger Bolz wrote,[2] "Fundamental to manufacturing engineering is the fact that it encompasses thorough 'before-the-fact' planning and detailed execution of the manufacturing process stages, not costly 'after-the-fact' correction of poorly planned or unplanned operations. The word 'engineering' implies just that—a broad carefully engineered study and development of the overall processing system, including a study of product suitability for economic processing." Therefore, the manufacturing process must be

---

[2] Roger Bolz, "Manufacturing Engineering," *Automation Magazine*, reprint, n.d.

treated as a system starting as early as possible in the design stage and planned through to the completed product.

Following is a basic outline by Bolz to show the typical functional duties found in successful manufacturing-engineering operations:

1. *Preliminary cost and process capability study.*
2. *Devise methods and facilities for manufacture.*
   Develop processes and facilities.
   Prepare drawings and specifications.
   Prepare manufacturing instructions.
3. *Provide manufacturing facilities.*
4. *Exercise engineering control of production.*
5. *Estimate costs.*
   Establish cost factors.
   Estimate preparation costs.
   Estimate manufacturing cost.
6. *Collaborate with design engineering.*
   Help develop new designs.
   Exploit new processes and materials.
   Exchange technical information.
7. *Other.*
   Assistance to shop.
   Production capacity analysis.
   Job rating.
   Patent activities.
   Industrial relations.

Research and Development

The shorter tooling and on-floor time possible with numerical control means that developmental items can be proved out in less time and at less cost than would be possible with conventional machines. The fact that even one part can be made economically on numerical control equipment means more research and development can be done with the direct assistance of manufacturing equipment. In fact, one large company is requiring the research and development function to generate all developmental parts by numerical control equipment. The philosophy behind this is that the parts for experimental work, development, pilot production, and full production can be achieved more rapidly, and more economically, with numerical control. Research and development is not restricted to one copy of the part as is often the case because of prohibitive costs in making, say, half a dozen copies with manual methods. Yet, when a developmental part is deemed satisfactory, the information for its manufacture is already on tape, which can be used directly or with minor modifications to manufacture the part.

Sales

Although numerical control does not normally imply organizational upheaval for the sales function, its merits in facilitating the sales job are substantial. The shorter lead time from idea to finished product made possible with numerical control means that sales activity may not be as rigidly restricted to standard off-the-shelf items. Rather, variations of product to broaden the sales scope are made more feasible in many cases. The common complaint of manufacturing that sales people sell "what we don't have" may not have such negative implications if there is an awareness of just what variations are more readily implemented via numerical control. This would be an oversimplification for complex assemblies, but, at the component level, numerical control indeed allows sales flexibility through diversification possibilities due to shorter lead times.

5

Selecting Equipment

PRODUCT REQUIREMENTS FOR NUMERICAL CONTROL

Before a decision to acquire numerical control equipment is finally made, a prospective purchaser should analyze his reasons for wanting it in the first place. Does he really need numerical control? Does his company have a long-range program for acquisition of capital equipment into which numerical control fits? Has he considered its "systems" character sufficiently to adequately prepare for it organizationally and procedurally? In particular, what is the type of work contemplated to be done with numerical control?

Generally, numerical control is considered most efficient when one or more of the following conditions prevails:

1. A diversity of parts is to be produced, so frequent changeovers of machine setup and a large tooling inventory are conventionally required. This situation is found often in the aerospace industries, where there is great variety in required parts, yet the quantities per type are usually not as large as is normally associated with, say, the automotive industry. This is not to say that only large plants satisfy this condition. Most job shops, small and large, owe their survival and success to this type of operation.

2. A new product is to be made, and there does not already exist conventional tooling for which the investment must be amortized. Of course, merely producing a new product is not

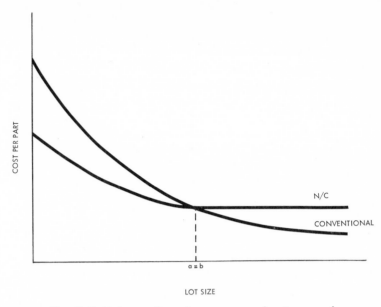

COST PER PART

N/C

CONVENTIONAL

a = b

LOT SIZE

Fig. 5–1. Unit manufacturing cost versus lot size, a = b.

sufficient cause for going to numerical control. However, new products particularly should prompt consideration of fabrication by numerical control.

3. Production lots are small or significantly variable in magnitude. Figure 5–1 gives a generalized picture of the unit manufacturing cost versus lot size for both conventional and numerical control fabrication. For conventional methods, the unit cost continuously decreases as lot size increases. Usually, unit cost for numerical control methods is less than that for conventional methods for small production runs. The numerical control unit cost also decreases for increasing lot size but at a lesser rate than for the conventional case. For a given set of conditions, there is a value of lot size, indicated in the diagram point as b, at which the unit cost achieves a value which remains essentially constant regardless of further increase in lot size. This occurs nominally at the point where certain initial costs have been written off. For a particular part, these costs usually comprise equipment investment and programming, tooling, and setup costs. In other words, except for amortization of initial costs (and, of course, material costs), unit cost for a

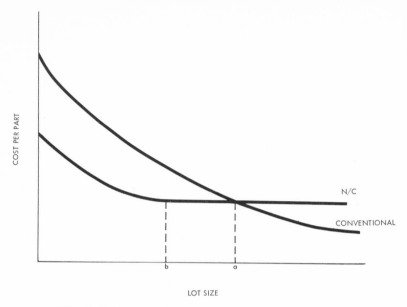

Fig. 5–2. Unit manufacturing cost versus lot size, $a > b$.

part made by numerical control remains approximately constant and insensitive to lot size. In the diagram, a represents a value corresponding to that lot size at which the two curves intersect so that unit cost by either method is nominally the same. For various parts and conditions, a may be greater than, or equal to b, so that the diagram of Fig. 5–2 or 5–3 may apply instead of Fig. 5–1.

4. Parts are of the same general class with respect to configuration but vary in dimensions. This situation is encountered particularly in the machine-tool industry, where there are families of parts similar in every respect except dimensions. A major advantage of numerical control is exploited in its facility for using the basic programming format of one part for all parts in the family.

5. Part dimensions and configurations cover such a broad range that adequate inventory of finished parts would be prohibitively large. Utilization of numerical control for manufacturing spare parts is an excellent example of how the capacity for providing spares can be retained without storing parts or tools and fixtures many years after a basic product is no longer

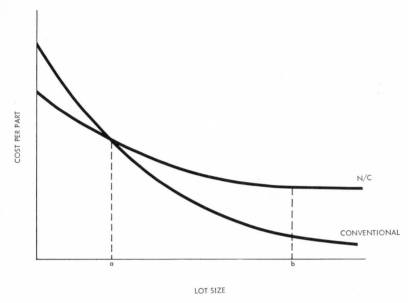

COST PER PART

N/C

CONVENTIONAL

a

b

LOT SIZE

Fig. 5–3. Unit manufacturing cost versus lot size, a < b.

marketed. Tapes corresponding to the parts are, of course, much more economical to store indefinitely than would be the actual parts or tooling.

Pipe bending by numerical control provides another illustration of the foregoing condition. Replacement of exhaust pipes for automobiles is not a rare occurrence. Yet, the bend configurations of pipes to accommodate the many automobile models are staggering in number. Supplying local installation centers with adequate inventory of different pipes to preclude delays when customers request a particular model is indeed prohibitive in cost. Too often, the only alternative is to require that the customer do advance ordering. This means that a large inventory of different pipes must be stored in a warehouse somewhere, since conventional production techniques preclude making each pipe to order. With numerical control, each configuration for every car model is represented by a tape, and only standard straight pipes bendable to any of the configurations (so that a much smaller number need be available) are stored.

6. The part configuration is complex, so relying upon manual implementation is difficult, especially if more than one (yet

a small quantity) is required. There are numerous examples of this type of requirement in the aerospace, military, and quasi-military areas where the number of copies required may be a half dozen or less. The part may have a curved shape, for instance, which can be described mathematically so that numerical control equipment can utilize its description for making the part. On the other hand, manual means with conventional equipment are not as readily amenable to making elaborate curves.

7. Parts involve manifold operations that are complicated, tedious, and hard to follow. In the first place, numerical control allows data on the tape to be utilized reliably again and again if satisfactorily programmed regardless of the intricacy of operations. The initial program would perhaps involve care and tedium, but, once it has been completed, any number of copies can be made with no repetition of that difficult phase of the job. With conventional equipment, the complexities involved in following instructions for the first part would continue to be involved undiminished for all copies. Secondly, unless the many required operations comprise repetitive patterns, usual conventional equipment utilizes essentially one tool at a time. On the other hand, numerical control machines can take advantage of a variety of tools in succession and interchangeably without penalty in time losses for machine stoppage, setups, transfers, etc.

8. Left- and right-hand parts are required. These parts are identical dimensionally and in every respect except that one is a "mirror image" of the other. For example, a housing with machining on both internal and external surfaces may be impossible to fabricate in one piece. For manufacturing purposes, it is made in two symmetrical halves, which are commonly called left- and right-hand parts. There are many instances of such symmetrical pieces in parts manufacturing. For the conventional case, this requirement entails going through every step in as much detail for a left-hand part as for a right-hand part, and vice versa. With numerical control, it is possible to utilize a feature that permits the fabrication of left- or right-hand parts with one tape programmed for either. In machines with this feature, actuation of a switch is all that is required to make a "mirror image" of the part for which the tape was programmed.

9. Parts are produced at intermittent times because of seasonal (or other cyclic) demand. There are seasonal influences, for example, in needs for many types of farm machinery. Numerical control precludes the necessity for costly storage of templates and other tools and fixtures. The versatility of numerical control equipment ensures its general usefulness in the "off" season for other work. Conventional equipment almost always includes some special equipment that would be of very little utility work other than that for which it is acquired.

Examples of Product Analysis

Figure 5–4 illustrates an example of a good candidate for numerical control. The part is a hydraulic manifold component whose life is relatively long with respect to its becoming obsolete. Demand for the part is expected to be continuous but not at a constant level. Consequently, it is made in runs that are not overly large. All surfaces require machining, so multiple-side exposures are necessary. It involves straight cut milling and contour milling on all surfaces, plus drilling of 17 holes, tapping of 13 holes, and counterboring of 4 holes. Three holes are counterbored and chamfered from the back. Additionally, accuracies and hole-to-hole tolerances are tight enough to make too many relocations from machine to machine (as conventional equipment would require) prohibitive.

This part would present numerous tooling and operational problems in the conventional case, not the least of which would be the long running time (exceeding eight hours) required for machining. It is made with a numerically controlled tool-changing machine tool wherewith *all* operations are performed in less than an hour. Two parts are loaded in a fixture. All appropriate operations are run on one side; then the part is turned to the opposite side and completed.

On the other hand, for the part shown in Fig. 5–5, numerical control fabrication presents both advantages and disadvantages. The part is made from an extrusion with a different series of holes required on each of five sides. In all, there are 124 holes to be drilled, 73 to be tapped, 24 to be counterbored, and 8 windows to be milled. Accuracy overall is to be to 0.001 inch, so essentially no accumulation of error is allowed. The job-lot requirements for this part run about 100 to 150 per month. When sales drop, part schedules must be extended accordingly.

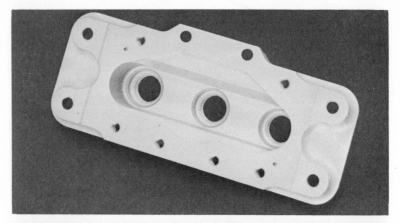

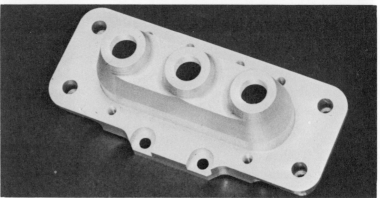

Fig. 5–4. A good candidate for numerical control.

Conventional machining presents difficulties, especially with regard to tooling. Costs would be high because of tight hole center tolerances and the close proximity of one hole to the other. In fact, adjacent holes are so close to each other that corresponding drill bushings in a single fixture are not feasible. At least two fixtures are necessary, which relate to each other through accurate tooling holes.

It is not surprising that the fabricator of this part decided upon a numerically controlled tool with an automatic indexing fixture. The part is drilled, tapped, and counterbored, and windows are milled in three different setups using the same index fixture. And the choice of numerical control is indeed good from the production standpoint, considering the difficulties of con-

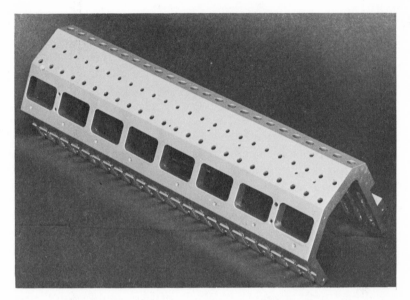

Fig. 5–5. Part with pros and cons for numerical control.

ventional tooling for this part. However, this is not to say that there are not some negative aspects to this choice. Mainly, the difficulty stems from the fact that this particular manufacturer does not have other work to put on the numerical control equipment. In other words, he has purchased flexibility that he cannot utilize.

It can be seen from this situation that the number of parts to be made, together with other requirements of the manufacturer, is a very critical factor in evaluating his choice. A steadier demand for the part under discussion, considering the lack of general work, might justify a special tool with capacity for simultaneous multiple-hole operations. The difficulty with expensive special machinery, however, is that it cannot be utilized for parts other than that for which it was originally designed. Yet, erratic demand would indeed result in equipment idle time.

Need for Thorough Analysis

It is important that the numerous factors discussed above be considered in evaluating the need for numerical control. Part characteristics may be amenable to efficient production by numerical control, but it is also necessary that other conditions exist to make the investment worthwhile. Often the investment is quite large. There is rarely one-to-one correspondence between conventional and numerical control machines. As mentioned before, one numerical control machine usually takes the place in function and work capacity of several conventional machines. By the same token, the numerical control machine is considerably more expensive than any of the individual simple machines it replaces. Therefore, the volume of work to be done on numerical control equipment should be sufficient to keep the machine relatively busy and help spread the high overhead rate to which expensive capital equipment contributes.

One manufacturer of fabricated components analyzed his requirements and found that the parts produced in his plant were compatible with numerical control techniques. He acquired three tape-controlled machining centers with tool-changing capability for milling, drilling, boring, tapping, reaming, counterboring, and chamfering. Each of the centers and their accessories cost from $150,000 to $300,000, so the total investment was over a half-million dollars. Production of the components was, indeed, efficient on these machines. In fact, it was so efficient that total weekly requirements were satisfied each week in three or four days. Having a half-million dollars of equipment sitting idle one or two days a week is, of course, extremely poor utilization of invested capital. In the particular plant under discussion, production personnel literally put "busy work" on the machines to avoid the image of idle equipment when visitors went through the area.

Efficient machine utilization is typical of what management looks for in evaluating new capital equipment. And, because the investment is large, management at a high level is concerned with the evaluation. Consequently, a thorough analysis of the economics involved beyond the mere utilization factor is necessary to justify acquisition of numerical control equipment.

TECHNICAL ASPECTS OF EQUIPMENT

If there are part requirements that suggest the need for numerical control, consideration must be given to technical aspects of the desired equipment. Unfortunately, the guidelines usually presented are for evaluating the controls only. Yet, a good control without an adequate machine tool is just as useless as a good machine tool without a proper control. Each must have certain minimum characteristics, but, even more important, the two must be suited to each other.

The prospective numerical control user must first of all make an analysis and evaluation of the type of work he intends to do with his equipment. He cannot simply decide that he wants to buy the biggest, fastest, and most versatile system available. In the first place, it is not practical (or even possible) to design a system where some parameters do not have to be compromised to get others. The biggest possible machine can hardly also be the fastest possible machine. Secondly, available equipment has practical limits of size, speed, and versatility dictated by the segments of the market they are intended to serve. Finally, as with any product, the more the features, the greater the cost.

After he has established the work scope, the prospective user must specify the vital statistics required of the machine system. He can do this in either of several ways. He may list the absolute minimum requirements that must be satisfied for his purposes and then proceed to search out available equipment meeting or exceeding these requirements. This has a drawback in that he may generate specifications for an overall system that simply does not exist and would be prohibitive in cost for a manufacturer to custom-build. He could, of course, attempt to match his requirements to available equipment and make compromises in parameters representing incompatibilities.

The most desirable approach would be for the prospective user to have some awareness of available systems so that his listing of requirements is made in conjunction with a study of latest machines on the market. In any event, a listing of parameters is mandatory if for no other purpose than to use the specifications in later computing potential cost savings for making his economic justification for purchase.

Machine-Tool Parameters

Type of Machine To Be Controlled. Metal-cutting tools available with numerical control include mills, drills, bores, grinders, lathes, profilers, planners, presses, shapers, and many combination machines. The prospective user may have work requirements for milling, drilling, and tapping. A machining center, for example, may do the job nicely. However, since such a system may cost anywhere from $75,000 to $500,000, he will have to look at his requirements more closely. He may find a drill, which with numerical control usually has capabilities for limited end milling and tapping and which may cost perhaps $30,000, will more than adequately accommodate his work. The available range of machine capabilities is quite large.

Table Machined Surface Size. Numerical control introduces little difference from the evaluation made for conventional tools in the consideration of machine-table size. Of course, this is dictated by the parts that must be handled. The carrying means must be large enough to hold the workpieces and sturdy enough to accommodate the power expended in cutting.

Number of Axes of Motion with Designations. There are machines for special jobs with only one or two movable axes. In most cases, however, where it is intended to do more general work, the machine to be used usually has three axes of motion. Perhaps it is not necessary that all three axes be under numerical control. For example, there are many turret drills with only two axes under control—those that move the part to desired locations. The third axis is the cutter axis, which may be under manual or simple sequential control.

There are also many cases in which the machine configuration may involve more than three axes of motion. In addition to the three customary orthogonal axes, a machine may have an index and/or rotary table representative of another axis. Further axes may be reflected in tilt of the table and head, for instance. Still another axis designation may be associated with relative motion between the spindle and its quill. The extension of the quill is usually called out as a separate axis, depending upon its degree of separate control.

Travel Range in Each Axis. The user must translate the knowledge about the type of work he intends to do and the parts on which he will operate to determine the travel requirements for the machine axes of motion. These travels are important in themselves but are necessary also to establish overall machine size.

Accuracy Required in Each Axis. It is vital that a prospective user be careful in setting accuracy requirements. In the first place, usual limitations on accuracy are placed by the machine tool rather than the control. It is unrealistic to expect that a machine with basic inaccuracies will somehow be corrected by simple application of a numerical control. In some cases, control means are available essentially to monitor machine performance in a particular axis and adapt corrective measures. One of the simple examples of this is the machine with lead-screw drive on the table in which the lead screw allows more "slop" than is desirable. The measuring means might be a linear scale whose feedback signals are more closely representative of the workpiece motion so that the control acts upon data from the scale and is relatively insensitive to the lead-screw vagaries. However, if the feedback and actuation means depend upon a mechanical linkage with no integrity, the control has no opportunity to make corrections.

One large manufacturer had a machine in which the table-to-ways tolerance was ±0.005 inch. Expectations of part accuracies to the control resolution of 0.0001 inch were in vain until the machine itself was mechanically corrected.

Present (and foreseeable future) technology indicates that accuracy bounds will probably be established by the machine tool. We know that, beyond a certain point, high accuracies in large machine tools comprising numbers of mating parts are extremely expensive to attain. Therefore, the prospective user is wise to realistically assess what he needs. One aerospace manufacturer found that his facilities people were purchasing almost all machines for part tolerances based on 0.0001 inch. Yet, the vast majority of operations performed required no better than 0.001 inch to 0.005 inch. To be sure, the manufacturer had need for machines to do 0.0001-inch work, but some selectivity would have saved him thousands of dollars in equipment costs. Of course, this is a general problem and is not unique to numerical control machines. However, these machines are individually so much more expensive than conventional tools that even greater care must be exercised in their selection.

Rapid Traverse in Each Axis. The user wants his machine to be "making chips" rather than expending time in motion during which there is no cutting. Therefore, when the machine elements move to or from a limit position, it is desirable that they move in the fastest mode possible—rapid traverse. This is a parameter most likely to be subject to the technical limitations of the drive components–control match. The user should give his preferences here but normally must accept the rapid traverse available on the machine system selected essentially on the basis of other parameters.

Many machines have rapid traverse rates of 150 to 200 inches per minute. It is possible technically with most machines to attain higher rates if higher-powered drives are designed for them. However, it must be remembered that attainment of rapid traverse (or stopping) from a given rate is not immediate. Deceleration periods to stop are sometimes much longer than the time savings that can be gained by making the rapid-traverse rate higher. Consequently, it is not always worth the extra complexity and cost to implement higher traverse rates.

Feed Rates and Ranges in Each Axis.   If the numerical control machine is to do milling, it is desirable to have a range of feed rates. These are either in steps from low to maximum feed in inches per minute or infinitely variable between the limits.

Number of Spindles.   The number of spindles takes on special significance with numerical control. In numerical control machines that do not have tool changers per se, the number of spindles is representative of tool-change capability. Tool-change machines may have 1 or 2 spindles with many cutters (20, 31, 60 or more) automatically selectable from a tool magazine. Simpler numerical control machines without tool changers are standardly available with 1, 2, 6, 8, or 20 spindles (in some cases, quills) each of which is automatically selectable.

Spindle-Speed Ranges.   Spindle-speed ranges are usually dependent on the type of spindle (mill, drill, bore, etc.) and the capacity (power) of the machine. These may range from a few to several thousand revolutions per minute. These too are usually available in numbers of steps or infinitely variable from the lowest to the highest.

Number of Tools.   If the machine is of the tool-changer type, the number of tools is important in establishing the extent of diverse work that can be performed per setup. Available tool changers accommodate 14 to 60 or more tools.

Cutter Speeds and Capacities. As with conventional tools, the user is interested in determining cutter sizes and capacities in the materials he will encounter. These are, of course, related to the spindle drive power.

Power Supplies. Unlike conventional tools, where the operator's muscle power supplies the necessary forces for moving the machine elements, numerical control machines are automatic in their drives. Therefore, it is important, both in terms of capacity and in terms of primary power, that specifications be established for drive motive means. These could be A.C. or D.C. motors, hydraulic drives, or the like.

Type of Auxiliary Functions. There is great variety in available auxiliary functions with numerical control machines. Almost any function that can be accommodated to actuation of an "on-off" switch is programmable with numerical control. One of the most popular is coolant-flow actuation, which is amenable to being turned on or off via auxiliary function control. The number of steps or ranges for each auxiliary function must also be considered.

Numerical Control Parameters

General numerical control specifications include parameters corresponding to those for the machine tool the functions of which are tape-controlled. For example, the number of axes of motion and the travel ranges in them are dictated by the machine specifications and do not have to be independently determined.

Type of Control, Programming, and Position-Feedback. The type of control—that is, positioning, contouring, or combination —is selected on the basis of the work to be done. Whether type of programming and position feedback are absolute or incremental is preordained by the make of control. The user of positioning controls may be influenced in his selection of make particularly by the convention used in his own drawings. For instance, base-line dimensioning may incline him toward absolute programming. On the other hand, most positioning controls use absolute programming, so he may accept it regardless of his own current blueprints.

Accuracy. Accuracy is of utmost importance to the user, but great care should be taken to distinguish between resolution and repeatability, two terms sometimes confusedly used interchangeably with accuracy. Accuracy is a measure of the numerical control system's ability to achieve a programmed location.

Resolution in Each Axis. Resolution is the smallest increment that can be programmed as input to the control. Controls are available with resolutions of 0.001, 0.002, 0.0005, and 0.0001 inch. In addition, there are controls for use in less accurate work with resolutions as coarse as 0.010 inch. There are, on the other hand, ultraprecise systems with resolutions of 0.000010 inch. However, most standard controls have resolutions of 0.001 inch or 0.0001 inch, with a growing tendency toward the latter.

Repeatability in Each Axis. Repeatability is the magnitude of error in repetitive deviations from a programmed location. Repeatabilities of 0.0002 inch are often achieved with 0.001-inch resolution systems. The total for a system is dependent upon the combination of the machine tool plus the control characteristics.

Type of Input Medium. The type of input for most controls is 1-inch-wide, eight-channel perforated tape.

Tape-reading Speed. Devices for interpreting data on the tape and translating them into acceptable input for the control are tape readers with speeds ranging usually from 20 characters per second to 1000 characters per second. Currently, most are 100 and 300 characters per second for positioning and 500 characters per second for contouring and combination controls.

Other Control Parameters. Other specifications either are selected straightforwardly in accordance with the machine specifications or come about as a consequence of characteristics of particular control makes. They include tape language and address system, number of auxiliary functions under control, options, operation modes (e.g., automatic, semiautomatic, manual), and power requirements.

Some of the specifications may not be applicable per se for the particular machine under consideration. For example, the part may not be carried on a "table," so another element's capacity may be more important. Nevertheless, the items can serve as a checklist suggesting the main elements specifications.

Requesting Quotations

After these parameters are established, the prospective user should document them so that subsequent communications with potential vendors will permit uniform requests for quotations. The most appropriate sources for system quotations are machine-tool companies that (1) produce total systems or (2) produce machines, but coordinate control requirements with manufacturers from whom they purchase controls. In either case, both responsibility for the marriage of machine tool and control and overall systems responsibility remain with the machine company. If the prospective user is not clear on this point, he should establish it as part of the requested quotation.

There are times when compatibility with existing equipment makes a particular control manufacturer's product preferable. In such a case, it is still recommended that quotations be requested from the machine-tool builder or builders with a distinct indication of control preference. Some machine-tool manufacturers will not consider controls other than their own. Others will, but may, for proper reasons, quote a price reflecting additional costs for marrying a particular control to their machine. And still others will, indeed, make their machines available with any of several makes of controls and most certainly will try to accommodate the preference. Most important of all is that a prospective user should not attempt to buy separately machine tools and controls without first establishing who will ensure compatible interfaces, who will perform the actual interconnections, and who is responsible for continuing service.

FITTING EXISTING EQUIPMENT WITH NUMERICAL CONTROL

It is possible that an evaluation of the machine requirements will result in specifications not unlike those for equipment already on hand. Consequently, the natural tendency is to consider the possibility of applying numerical control to the existing conventional machine, that is, "retroactively fitting" or "retrofitting." Several control manufacturers produce equipment primarily intended for retrofit and have applications engineers assigned specifically for retrofit. Additionally, machine-building companies are increasing their attention to converting older machines to numerical control. The main reason for this retrofit activity is the interest in salvaging the large investment represented by the many existing machines that would be many times more productive if numerically controlled.

Although there are instances where retrofit has been accomplished quite successfully, the results have not always been gratifying, and the number of retrofitted machines is very low relative to the number of new numerical control installations.

The major obstacle standing in the way of wider applications of retrofitted controls is the cost of the job engineering required on the existing mechanical portions to adapt them to the new controls.[1] Machines to be retrofitted were designed before numerical controls were available and, in very many cases, before the numerical control technology was conceived. These machines reflect no thought of future possible accommodation of control elements, so drive and positioning members are not readily accessible for conversion. Newer conventional machines being introduced are more amenable to the requirements of economical retrofitting.

Important, too, is the fact that retrofitted systems seldom approach the performance levels of new integrated systems.[2] In the first place, machines designed specifically for tape control often incorporate new features that cannot be easily duplicated in the retrofit except through costly modifications. Also, in many cases, a used machine is apt to be most worn in such areas as the ways, slides, gears, drives, and other basic points of accuracy, which are the very elements the user must automatically control. Therefore, retrofitting for complete conversion of a machine to numerical controls will probably not make a significant impact until the new generation of machines is phased in.

---

[1] Charles Weiner, "Which Door to Tape Control?" with comment by William C. Leone, *Tooling and Production*, January, 1961, p. 50.

[2] William C. Leone, "Machine Productivity Renewed," *Machinery*, February, 1964, pp. 116–18.

Position Displays

Another aspect of retrofitting has been partial conversion of conventional machines—this partial conversion being the adaptation of position indicators and readout devices, which give some of the benefits of numerical controls but still require manual (including push-button) control of the drives. The application to manually controlled machines (including retrofit to existing machines) allows the operator to position the machines more rapidly and with greater accuracy, thereby reducing operator fatigue and avoiding the errors associated with the reading of dials and verniers.

The position-indicating device is used in conjunction with a position transducer geared to a lead screw or precision rack. Any means of mechanical coupling is acceptable which provides for a transducer output accurately proportional to the relative motion of the machine members. Position information is usually displayed in numerical form.

In use, the operator sets the display at zero at a suitable reference point. For example, this reference point may be at one end of the machine slide or the center line of the part. As the machine slide moves away from this zero point, a counter follows the transducer signals and continuously displays the absolute position. If the slide moves back toward zero, the counter reverses or subtracts and still displays the absolute position from zero. In addition to displaying dimensional data, the display incorporates means to show automatically the position of the machine slide with reference to the zero point. If the reference point is on the part with machine operations on both sides of the zero point (such as the center line of a bolt circle), the counter subtracts as the tool moves toward the zero point. As it passes through zero, the counter automatically switches and starts adding. At the same instant, the direction indication changes.

The counter can be reset to zero after each operation if the operator elects to do so. In this manner, counting starts from zero and the incremental dimension of slide movement is displayed. Displays also sometimes incorporate "dial-in" capability. This allows one to enter reference numbers after selecting a desired zero position and add to (or subtract from) them as the machine element moves.

This type of readout allows both small and large shops to avail themselves of partial automation where particular machine considerations do not warrant full numerical control.

An area of considerable interest for users of position displays is in association with inspection devices. The increase of automatic production machinery now is producing parts at an accelerated rate—so fast, in fact, that traditional checking means are not always economical in time and money. In some instances, 100 per cent inspection is required. The result is that the gains from mechanized production are almost completely canceled by the bottleneck at inspection. Methods utilizing the numerical readouts have improved the efficiency of checking to a point where inspection departments can better keep pace with production output.

ECONOMICS OF JUSTIFICATION

The justification procedures for acquiring numerical control vary greatly, depending upon the company doing the buying. Most companies have established guidelines for justifying capital expenditures. However, some are very elaborate and detailed tabulations with consideration of effects on all company direct and indirect cost factors. Others are quite simple, tantamount to purchase based on the whims of someone in management. Because some of the justification computations are unique to numerical control and, therefore, may differ from what management has been accustomed to use for purposes of evaluation, various equipment manufacturers have developed guidelines themselves for buyers to use. The result is that many different procedures exist, any of which the prospective buyer can adopt or modify to suit his own situation.

One Year's Production Method

One medium-size manufacturer bases his evaluations of machinery on the cost of one year's production for several "typical" parts for which he requires equipment. His formula for the cost of one year's production is

Cost of setup
+ Cost of loading and unloading
+ Cost of machining
+ Cost of transporting
+ Cost of tooling
+ Cost of planning
+ Cost of amortization
+ Cost of work-in-process inventory for one year's production at machine production rate.

Detailed assumptions are made for the number of parts in a lot release run on one setup (assumed constant), the number of lots of one kind of part in a year, the number of different kinds of parts made in one year, and realization factors. For the amortization, he makes estimates or computes yearly depreciation based on the original cost and life of the prospective machine, floor-space cost per year, maintenance cost per year, property tax per year, insurance cost per year, and power cost per year. He calculates the cost of work-in-process inventory based on estimates of interest rates, the average value of one part before machining, the time required to machine one part, the average value of machining cost on one part, and the wasted time that elapses before the part is assembled and put into finished stores.

The formula for cost of one year's production is made up of the particular cost segments indicated so that it can be readily applied to both conventional and numerical control machines. Comparison between any two sets of machines is not difficult.

Return-on-Investment Method—MAPI

Many of the larger manufacturers use procedures that follow MAPI [3] guidelines. There is some variation among the users in application of the MAPI format, but, in general, the procedures consider

1. Required investment
2. Effect of project on operating costs
3. Non-operating advantage from project
4. Chart allowance for depreciation
5. First-year percentage return on investment

[3] George Terborgh, *Business Investment Policy: A MAPI Study and Manual* (Washington, D.C.: Machinery and Allied Products Institute and Council for Technological Advancement).

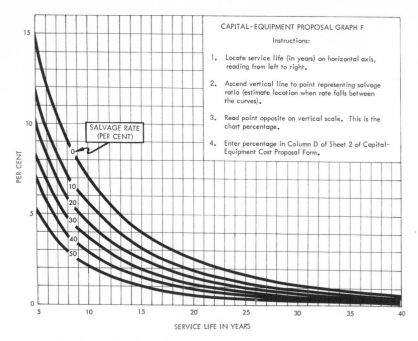

Fig. 5–6. Graph used for capital-equipment cost proposals.

The following pages are an example of the justification format used. In the chart (Fig. 5–6) relating salvage rate (sometimes called ratio) and service life, the percentage values increase vertically in increments of 1 per cent from 0 per cent at the bottom to 16 per cent at the top. The service-life values increase horizontally in increments of 1 year from 5 years at the left to 40 years at the right.

In order to better appreciate the application of the MAPI-type procedure for equipment justification, an actual example of a cost proposal made (and accepted) in one company is given. All of the pertinent numerical data in the following are exactly as presented to the company's management.

The company in this case is a manufacturer of components used in jet engines. The cost proposal was submitted to justify the replacement of three drill presses, two milling machines, and one turret lathe ranging in age from ten to twenty-five years with a numerically controlled turret drill. The latter was to have eight spindles in the turret and was to be controlled in three axes.

	SIX CONVENTIONAL MACHINES (′)			NUMERICAL CONTROL MACHINE (″)			
	1	2	3	4	5	6	7
Part No.	Tooling Estimate ($)	Setup Hrs.	Machine Time (hrs./part)	Added Programming ($)	Tooling Estimate ($)	Setup Hrs.	Machine Time (hrs./part)
FC611	2010	30.2	0.802	650	430	6.3	0.670
B201	650	7.3	0.755	550	175	3.0	0.216
TC112	3050	41.0	0.646	1250	730	11.7	0.274
J332	2300	25.3	1.430	1350	545	9.2	1.373
SP776	1875	32.7	0.518	750	900	8.3	0.200
e = Total	9885	136.5	4.151	4550	2780	38.5	2.733
f = Average/part	1977	27.3	0.830	910	556	7.7	0.547

Fig. 5–7. Justification worksheet.

Five representative parts were selected by the manufacturer to be studied in some detail to compare the more important cost parameters in conventional and numerical control machining using the existing and proposed tools. In this case, five parts were selected as the basis for study. It was felt that the manufacture of these five alone would justify the machine replacement. In some cases, the manufacturer may use many more parts, 20 or 30, on which he bases his comparisons. Or he may use as few as 4 or 5, merely representative of a much higher number he intends to make with the new equipment. There is no fixed rule as to the number of parts to be considered. The prospective user must use his best judgment as to the kinds of parts that will actually be made with the new equipment. The validity of his justification is no greater than the similarity between the parts selected for making the justification and those ultimately produced with the equipment.

It is also important to note that, although this example treats a situation in which several conventional tools are to be replaced with a numerical control machine, use of the MAPI format is not restricted to replacement (nor to numerical con-

PURCHASE REQUISITION for CAPITAL EQUIPMENT				WORK SHEET	
Date	Brass Tag No.	Account	Job. No.		

Suggested Vendor:	Date of his quotation:
Mfr. A	

DESCRIPTION		PRICE
(1) Model xxx	Mfr. A	Type Vertical Turret Drill
with 3 - axis numerical positioning control. Complete as per marked quotation except N/C should be		
re-quoted for XYZ control adaptation incorporating latest design improvements at time of delivery.		

Voltage wanted: (Circle One)	Color Paint:		Total Value of
	7B-Std. Light	OR:	Order
110 (220) 440 ____	_X_ Mach. Tool Gray		
Delivery Quoted	Delivery Required	Deliver to Station	$49,745.00
90 - 120 days	A.S.A.P.	#31	
Plant where equipment will be used:	Department	Supervisor in charge of using equipment	
Plant #2	# 4-18		

OTHER MAKES YOU CONSIDERED	DATE OF QUOTE	DELIVERY	PRICE
Manufacturer B	-	-	-
Manufacturer C	-	-	-

What decided selection:	____ Lowest Bid	____ Early Delivery	_X_ Past Experience
_X_ Features	_X_ ▓▓ Standardized	____ Quality	____ Only Available Source

Did you check other departments or plants for similar unused facilities or equipment?	If Yes - Who did you contact?
_X_ Yes ____ No	-

Is this additional equipment? ____ Yes _X_ No	Approx. age of replaced equipt.	Will new equipment involve rearrangement?
or		
Does this replace obsolete equipment? _X_ Yes ____ No	10 to 25 yrs.	_X_ Yes ____ No
What disposition planned for replaced equipment?	Products on which this equipment will be used:	
Salvage	General machine work - Dept. #4 - 18	

Why is this equipment needed?
To update present facilities and to start N/C operations in Dept. #4 - 18.   In addition, this machine is
to help in reducing the continued need for overtime and farming out in Dept. #4 - 18.

Requisitioned by:	Purchase Approval Signatures:
Supervisors Ok.	

Fig. 5–8. Worksheet–purchase requisition for capital equipment.

trol machine acquisition). It is one general procedure for standardizing the justification approach for capital equipment such as machine tools.

In the example, the study on the five representative parts included the gathering and computation of estimated data on hours required for setup, tooling cost, machining time per part, and conventional versus numerical control programming cost differential. For the conventional case, the figures represent totals for all six conventional machines. In each case, the programming cost differential represents the added cost for complete planning and preparation of a tape for numerical control over the planning details for the six conventional machines. The data shown in Fig. 5–7 were compiled on these parts.

Following are the actual analysis sheets filled out with the appropriate information relating to the machines and the parts for which the foregoing data were accumulated. The worksheet (Fig. 5–8) is self-explanatory. It lists the overall vital statistics and cites other makes considered in presenting the recommendation to management. Immediately following this page is the MAPI format (Fig. 5–9). The items will be discussed briefly in the order (and with the same numbering) corresponding to their listing on the MAPI format. In those instances where blank spaces are not filled in, one or more of the reasons that may apply are

1. The item does not represent a sufficiently significant magnitude of difference between numerical control and conventional
2. The item is not applicable
3. The item is an unknown or quite difficult to estimate but is believed to favor the justification; yet the other items are believed to be adequate to warrant the recommendation for equipment acquisition

I. REQUIRED INVESTMENT. Except where noted (line 21) it is better to make an intelligent estimate on each item than to leave it blank. The originators of the proposal are usually the best informed and able to make the best estimate.

Items 1–7. These items in the proposal are self-explanatory.

II. NEXT-YEAR ADVANTAGE FROM PROJECT. Use first year of project operation. For projects with a significant breakin period, use performance after breakin.

Item 8. Operations were assumed on a basis of two 8-hour shifts. Considering five-day weeks and allowing for six holidays, the operating rate is 4064 hours per year.

Item 9. Direct labor corresponds to the machining hours. For the numerical control machine, the average machining time per part is 0.547 hour. Therefore, in 4064 hours, assuming 100 per cent utilization of available machine time, 7429 parts can be made.

DESCRIPTION: *Replace 3 drills, 2 mills, 1 lathe with N/c turret drill*

I. Required Investment

1	Installed cost of project (include freight, foundations, installation).	$ *49,745*		1
2	Less *7*% investment tax credit.	*3,482*		2
3	Net installed cost of project. (line 1 minus line 2).		$ *46,263*	3
4	Salvage value of old equipment.	*2,000*		4
5	Cost to rebuild old equipment or replace part of a line.	—		5
6	TOTAL of line 4 & line 5 credited against cost of new project.		*2,000*	6
7	Net investment required (line 3 minus line 6).		*44,263*	7

II. Next-Year Advantage from Project

A. Operating Advantage

8	Assumed operating rate or project (hours per year).		*4,064*	8

EFFECT OF PROJECT ON OPERATING COSTS

		INCREASE	DECREASE	
9	Direct labor (include overtime percentage for the product).	$	$ *6,097*	9
10	Fringe benefits (divisional average for previous year).		*2,208*	10
11	Maintenance (include only variance in ordinary maintenance).	*600*		11
12	Tooling (enter only if voids or adds expense such as a. Drill jigs voided b. Reduced tool expense due to greater rigidity, etc.).		*7,105*	12
13	Materials and supplies (example—abrasive belts, etc.).	—	—	13
14	Scrap and rework (enter the estimated increase or decrease).	—	—	14

Fig. 5–9. Cost analysis for replacing conventional machines with a numerical control machine.

COMPANY CAPITAL-EQUIPMENT COST PROPOSAL (cont'd)

		INCREASE	DECREASE	
15	Downtime (examples of causes of greater downtime: a. Worn equipment b. Complex new equipment, i.e., electronic & tape controls, etc.).	$ —	$ —	15
16	Power (increased or decreased hp)	—	—	16
17	Floor space (enter only a major item of increase or decrease).			17
18	Property taxes & insurance (you must always enter this figure).	895		18
19	Subcontracting (enter savings by mfg. within the plant, i.e., handling, high cost at outside source, etc.).			19
20	Inventory (reduced setup cost and lead time resulting in smaller lots and thereby reducing carrying cost, shop and crib floor space).		5,000	20
21	Safety (enter only if voids a known risk).			21
22	Flexibility (enter only in an outstanding instance).			22
23	Other (explain or attach explanation of any amount entered).	4,550		23
24	TOTAL.	6,045 A	20,410 B	24
25	Next-year operating advantage (line 24B minus line 24A).		14,365	25

B. Non-operating Advantage from Project
(Use only if there is an entry in line 6.)

26	A. Decline of salvage value of old equipment next year.	$ 0 A
	B. Next-year allocation of rebuild cost, etc. (line 5 divided by the number of years the rebuilt or new equipment will add to the life of the old equipment or line).	— B
	TOTAL (Line 26A plus line 26B).	$ 0 26

Fig. 5–9 *(continued).*

141

COMPANY CAPITAL-EQUIPMENT COST PROPOSAL (cont'd)

C. Total Advantage

27   Total next-year advantage from project
(line 25 plus line 26). $ _14,365_ 27

III. Computation of Urgency Rating

28   Total next-year advantage after income tax
(line 27 minus income tax @ 48% rate).   _7,470_ 28

29   MAPI chart allowance for project (total of
Column E below).   _2,958_ 29

        Column B—Estimated Service Life—use 10
years unless analyst can fully justify less
or more; then attach explanation.
Column C—Estimated Terminal Salvage—
use zero except in unusual cases; then
attach full explanation.
Column D—Percentage—use standard
graph F (Fig. 5-6) except in unusual
cases (attach explanation).

Item or Group	Net Installed Cost of Item or Group (Line 3)	Estimated Service Life (years)	Estimated Terminal Salvage (per cent of cost)	Percentage from Graph F	Chart Percentage ×Cost (D × A)
A	B	C	D	E	
	42,263	10	0	7 %	2,958
			•		

                                 TOTAL $ _2,958_

30   Amount available for return on investment
(line 28 minus line 29). $ _4,512_ 30

31   Urgency rating (line 30 divided by line 7) =
first-year percentage return on investment.   _10.2%_ 31

Fig. 5—9 *(concluded).*

This same number of parts would require (7429 × 0.830) machining hours to make with the six conventional tools, since the average machining time per part on them is 0.830 hour. Consequently, the direct-labor decrease with numerical control is 7429 (0.830 − 0.547) times the hourly direct rate. Rates vary widely depending upon area, classification structure, etc., but that for the company in question and used in standards determinations is $2.90 per hour. The direct-labor computation results in a $6097 decrease for the year, using numerical control for comparable output.

Item 10. Just what constitutes fringe benefits is a function of company or, sometimes, union policies and varies very widely according to different financial-department treatments. In some companies, fringes are never segregated and are always part of the overall manufacturing overhead. In others, they are, indeed, looked upon as a separate factor, which may include costs per man-hour for company contributions to such things as pensions, insurance, and deferred pay compensation. For the company under discussion, fringes are assumed at $1.05 per man-hour. Item 10 is computed by multiplying the net difference in direct man-hours for numerical control versus conventional machining by the $1.05 rate, that is, 7429 × (0.830 − 0.547) × $1.05, equal to $2208. Again the result favors numerical control and is listed as a cost decrease.

Item 11. Ordinary maintenance is required for both numerical control and conventional machines. However, the approach was conservative in justifying the new equipment so that a major portion of a preventive-maintenance service contract for the numerical control machine was allocated to cost increase. In many cases, a cost decrease is possible with numerical control, especially when it replaces worn equipment.

Item 12. Since the company was basing its study on the five parts mentioned before, the tooling cost difference is $9885 (conventional tooling cost) minus $2780 (numerical control tooling cost), or a decrease of $7105. If more parts are considered, the tooling-cost difference is the difference of the average tooling cost per part multiplied by the number of *different* parts assumed to be made during the year.

Items 13–17. These are left blank in accordance with previously mentioned reasons.

Item 18. Taxes and insurance factors are dependent upon local situations. In this case, a factor of approximately 1.8 per cent times acquisition cost is assumed.

Item 19. This was believed definitely to favor the numerical control case but is usually not easily estimated. In this case, the people presenting the proposal felt that other items would be sufficient to result in a favorable recommendation.

Item 20. This is another item usually heavily favoring numerical control. An estimate of $5000 cost decrease was made. This is conservative in view of the fact that setup alone represents average savings of (27.3 − 7.7), or 19.6, hours per part with numerical control. If it is assumed that the average lot size is 100 parts, the 7429 parts could require over 74 setups. Therefore, on this basis, setup represents a cost decrease of 74 × 19.6 man-hours. If the $2.90 rate is again applied, this portion alone of cost decrease is over $4000.

Items 21–22. These items were not considered in detail.

Item 23. The programming cost differential was a large enough item to be considered. Since numerical control involved more time for the complete programming, the differential of $4550 is a cost increase.

Items 24–25. These items are self-explanatory.

Item 26. Since the salvage value of the old equipment was so low, no decline was assumed for the year.

Items 27–28. These are self-explanatory.

III. Computation of Urgency Rating. This is a computation for the first-year return on investment. Its value determines whether purchase of new capital equipment can be economically justified.

Item 29. This results from filling out the columnar information below Item 29.

A is self-explanatory.

B is estimated as ten years. This depends upon the equipment and local fiscal policies.

C is assumed to be zero, which represents an ultraconservative approach. In other words, the proposal shows justification even if such an expensive piece of equipment lasts only ten years and is worth nothing at the end of ten years.

D results from applying B and C to the graph. Incidentally, the graph is essentially a set of depreciation curves based on double-rate declining balance or sum-of-the-digits methods. There are corresponding MAPI chart curves for straight-line tax depreciation.

E results from multiplying D × A.

Items 30 and 31 follow automatically as shown on the proposal sheet.

It can be seen that this example yielded an "urgency rating" of 10.2 per cent. For all practical purposes, this signifies a return on investment of 10.2 per cent. Whether this is a good return depends upon what the investor can reasonably expect to earn with his money via other avenues of investment. Justifications for numerical control equipment have been presented exhibiting returns from a few per cent to over 100 per cent, depending upon the assumptions made. Some managements would accept a proposal based on intangibles alone and would not be unduly influenced because a return of only 1 or 2 per cent could be shown. On the other hand, some managements may insist on 20 per cent or more for the urgency rating. However, in most manufacturing situations, 8 per cent or more is quite favorable, so 10.2 per cent would be considered very good indeed, especially when one considers that it was derived using conservative estimates.

Simplified Worksheet for Economic Studies

Following is a typical justification procedure recommended by an equipment manufacturer to aid a prospective user in making his evaluation. It is extremely simple because it does not detail various indirect cost factors that are ordinarily considered by accounting people but are somewhat foreign to some of the operating personnel who participate in the recommendation and choice of equipment. Rather, rules of thumb are used for computing the principal cost factors. The following procedure is followed for obtaining one measure of "productivity ratio":

	Conventional	Numerical Control

$g = \begin{cases} \text{average number} \\ \text{of parts per pro-} \\ \text{duction order} \end{cases}$ $g' = g$ $= \underline{\quad}$ $g'' = g$ $= \underline{\quad}$

$h = \begin{cases} \text{total machine} \\ \text{hours per pro-} \\ \text{duction order} \end{cases}$ $h' = g \times f_3 = \underline{\quad}$hrs. $h'' = g \times f_7 = \underline{\quad}$hrs.

$k = \begin{cases} \text{total hours} \\ \text{(machine + set-} \\ \text{up)} \end{cases}$ $k' = h' + f_2 = \underline{\quad}$hrs. $k'' = h'' + f_6 = \underline{\quad}$hrs.

$m = \begin{cases} \text{productivity} \\ \text{ratio} \end{cases} = k'/k''$

$$= \frac{(g \times f_3) + f_2}{(g \times f_7) + f_6} = \underline{\quad}$$

		Conventional ($'$)	Numerical Control ($''$)
		(2032 × shifts)	(A'× 1/m)
A	Machine utilization (hrs.)	_4,064_ hrs.	_2,296_ hrs.
B	Direct labor ($2.90/hr.)× A	$ _11,786_	$ _6,658_
C	Indirect labor ($2.90/hr.)× A	$ _4,267_	$ _2,411_
D	Scrap/rework ($) 10% of (line B plus line C)	$ _1,605_	$ _907_
E	Tooling ($) ($f_1$)	$ _1,977_ (f_5)	$ _556_
F	Added programming	$ _____	$ _910_
G	Total running cost (lines B through F)	$ _19,635_	$ _11,442_
H	Net operating cost favoring numerical control (line G' minus line G″)		$ _8,193_
I	Acquisition cost of numerical control system		$ _49,745_
J	Installation, transportation		$ _included in I_
K	Net investment		$ _49,745_
L	Numerical control amortization period $\frac{K}{H} \times 12 =$ _72.8_ months		
M	Numerical control service life = _10_ years		
N	Yearly investment, $\frac{K}{M}$		$ _4,974_
O	Next year's savings (line H minus line N)		$ _3,219_

Fig. 5–10. Simplified worksheet for calculating next year's savings.

This productivity ratio is used in the "simplified worksheet for calculating next year's savings" (Fig. 5–10), which assumes that maintenance, power, labor required on the parts on other machines besides those being compared, and inspection are essentially a "wash." This is based on the fact that, although maintenance and power costs per hour are higher for numerical control than for conventional tools, the comparable time for the same production is normally shorter for numerical control, which more than offsets the differential in rates. The items assumed equal in total would favor numerical control in most cases, so this analysis assumes a justification procedure based on the more dramatically affected cost items. Scrap and rework are taken simply as 10 per cent of direct plus indirect labor costs for each; again, numerical control most often should certainly reflect a figure very much less than the 10 per cent so that, in this respect also, the simplified worksheet reflects a conservative approach.

It may be useful to follow the steps in this approach for the parts considered before relative to the MAPI-type justification. Assuming that the average number of parts per production order is 100,

$$g' = g'' = 100$$

The average machine hours per part are 0.830 and 0.547 for conventional and numerical control, respectively. Therefore, per production order, the total machine hours are 100 times these hours.

$$h' = 83$$
$$h'' = 54.7$$

Total hours (machine + setup) are represented by the symbol k.

$$k' = 83 + 27.3 = 110.3$$
$$k'' = 54.7 + 7.7 = 62.4$$

The productivity ratio is

$$m = \frac{k'}{k''} = \frac{110.3}{62.4} = 1.77$$

It can be seen that the result is not identical to that using the MAPI approach. However, the result is similar in that it is attractive enough in this case to warrant purchase of the numerical control equipment. Yet, the checklist procedure is extremely simple to apply.

Investment Consideration

In all three of the foregoing justification procedures, and in most others that are frequently used, the basis of evaluation is a comparison of conventional versus numerical control costs to determine whether savings (and, therefore, a measure of return on investment) are large enough to warrant purchase of the numerical control equipment. It must be noted that many aspects of capital-equipment acquisition reach beyond these justification procedures. The part the new equipment plays with regard to overall plant capacity and its compatibility in the "manufacturing system" must, of course, be carefully evaluated. From a fiscal point of view, consideration must be given to such things as the discounted cash value of the funds that would have to be invested in new equipment.

An interesting observation on financial aspects of numerical control was indicated in an SRI report on numerical control: "An increase in fixed assets (resulting from installation of costly numerical control equipment), of course, raises capital requirements, especially for long-term funded debt and equity financing . . . Adoption of numerical control programs sometimes makes possible reductions in inventory. To financial institutions this may mean a switch from demand for short-term to long-term debt."[4] This is indeed a point for prospective users to ponder—cost of more elaborate equipment will increase long-term investment; lower inventory requirements with numerical control will reduce short-term investment.

PITFALLS

In its first several years of life, numerical control equipment did not encounter an entirely favorable customer reception. This can be attributed to a number of causes.

The first systems of any significance put to actual use were of the contouring type and were built under United States Air Force sponsorship. They were necessarily somewhat complex and, therefore, costly. Aircraft work intended for these first systems was far from typical of the machining work in the traditional industries that utilize metalworking machines. The operations looked very elaborate. Furthermore, programming for path control, especially at a time when only tedious techniques were available, seemed a complete mystery to all but the controls designers themselves. Unfortunately, then, much of the interested public associated numerical control with con-

    [4] "Numerical Control," Long Range Planning Report No. 91, Stanford Research Institute, Menlo Park, Calif.

touring only and concluded it was overcomplicated and too expensive for the more general applications.

As commercial models entered the scene, there was much sales promotion and publicity in the trade journals. Well-meaning propaganda on the merits of numerical control overwhelmed an eager but inexperienced user public. Many forward-thinking plant managers decided to venture into numerical control too early and were not adequately prepared for it. They were led to expect virtual miracles by the excessive enthusiasm of the equipment manufacturers. Yet, in many cases, the equipment justifications had been ill conceived, so results were, indeed, disappointing. Although very many of the problems were directly attributable to misapplication of the equipment, the first users were disillusioned and often directed the blame to numerical control in general.

Projections of the market for numerically controlled machines have always indicated a bright future for numerical control. In metal cutting alone, the promise of early acceptance and rapidly expanding usage led to forecasts that 50 to 75 per cent of all tools would be controlled by numerical control within a few years of the first systems' appearance. This attracted a large number of control manufacturers to a seemingly lucrative new industry, and the competitive race to put equipment on the market was joined. Unfortunately, much of the equipment gave the numerical control concept another "black eye."

In the first place, numerical controls represent fairly sophisticated electronic systems comprising many components. Inherent characteristics of the types of components normally associated with these systems result in some failure or downtime. Although the failure rate by ordinary standards may have been quite reasonable in some of the first controls, the numbers of components in numerical control systems are high, so overall control downtime seemed excessive in terms of the day-in-and-day-out machine performance expected. Additionally, competition in controls has prompted the use of techniques and components still in their development stage. In some cases, the equipment put on the market has been essentially developmental considering that some of these advanced concepts were statistically untried in a production sense. Add to all this the

equipment introduced into the market that was poorly designed
and implemented. An impression was created, therefore, that
numerical control equipment could not be made to operate
without prohibitive downtime—that it was inherently unre-
liable.

The controls of numerical control are supplied by either
controls manufacturers or departmental entities within machine-tool companies. From a system point of view, this has
generated problems. There has been a basic tendency for machines and controls to be designed and built separately,
creating problems of interconnections between the two after
each is fundamentally completed and incompatibilities already
built in. Fortunately, a realization of this problem has generated
designs of machines and controls tailored for each other so that
truly integrated systems have emerged with very favorable
results. In general, manufacturers and users alike have made
good use of their experiences so that most of the errors of the
past are slowly but surely being corrected or avoided.

Problems Encountered with Numerical Control

A survey made by the *American Machinist* lists the biggest
user problems currently encountered with numerical control.[5]
A summary of the listing follows in terms of the percentage of
plants reporting. (The total exceeds 100 per cent because of
multiple answers.)

Problem Category	Per Cent of Plants
Electrical and electronic	32.4
Maintenance in general	16.7
Learning, attitude, personnel	12.0
Programs, tapes	9.7
Tolerances, repeat, inspect	8.3
Mechanical difficulties	8.3
Utilization, scheduling	6.0
Tooling, setup	5.6
Installations, debugging	2.8
Have no problems with numerical control	6.0
No answer	10.2

[5] R. L. Hatschek, "NC Today," *American Machinist*, Special Report No. 579,
November 22, 1965.

A survey of builders and users resulted in the following list of causes for downtime on numerical control equipment:

Inadequate planning	Machine breakdown: waiting for service
Wait for material	Under repairs
Programming error	Preventive maintenance
Unplanned deviation from tape	Planned modification
Insufficient setup information	Chip cleanup
Troubles with fixturing	Waiting for inspection
Tape tryout	Inspection
Cutting tools not available	Waiting for work
Gages not available	Demonstrations
Fixtures not available	Part design
Tryout: machine warmup	Quality control engineering[6]

The pitfalls suggested by the surveys confirm the need for a sound program of maintenance and service for both tape controls and machine tools.

TRAINING OF SERVICE PERSONNEL

By way of example, a course outline developed by a manufacturer of controls follows. The course is for training of service personnel on a particular make of numerical positioning controls. The course is oriented toward the turret drilling machine, since this has been one of the predominant applications for the control. However, the program of training for other machine applications follows the outline in every detail except those that refer specifically to drilling-machine functions. Appropriate modifications are made for special machines.

TRAINING OUTLINE FOR NUMERICAL POSITIONING CONTROLS

    I. Brief Description of Controls
        A. Standard equipment capability
            1. Position display
            2. Photoelectric reader
            3. Mill function
            4. Tapping function
            5. Coolant

[6] R. H. Eshelman, "Tomorrow's Tool Needs—Today," *Iron Age*, June 10, 1965, p. 143.

      6. Feeds and speeds for vertical axis

      7. Selective dwell

  B. Optional equipment capability

      1. Tool offset

      2. Parity check

      3. Programmed mill feed rates

      4. Tape spooler

      5. Air blast

      6. Rotary table

      7. Sequence number

II. Programming

  A. Relationship of part, program, tape, and controls

      1. Incremental and absolute programming

        a. Zero offset

        b. Multiple part program

        c. Round number increments

        d. Program check by addition of plus and minus numbers

      2. Part location

        a. With subplate (show subplate and origin point)

        b. Without subplate (explain with tooling hole)

      3. Types of tape format

        a. Tab sequential

        b. Word address

        c. Fixed block

  B. Tape format

      1. How to program dimensional information

      2. Preparatory commands—Z-axis—why and how used

      3. Auxiliary functions—feeds, speeds, etc.

      4. Miscellaneous functions

  C. Explanation of BCD punch number system

      1. Reading information directly from tape—mandatory for maintenance personnel

  D. Tape for sample test part

III. Equipment Operation (demonstrated with control as much as possible)

  A. Function of control panel switches and indicators

      1. When each is used in dial, block, and automatic modes

      2. Relationship of each switch to others and machine operation

  B. Machine control panel switches and functions

      1. Dependence upon machine type

      2. Typical type as example

C. Manual operation of machine and control
  1. Use of position display for measuring
  2. Tool offset procedure showing use of 2-axis display
D. Dial operation of machine and control
  1. All necessary switch settings for machine and control
  2. Use of this mode for first part location of single part fabrication
E. Block mode
  1. All necessary switch settings for machine and control
  2. Use of this mode for checking out tapes
F. Auto mode
  1. All necessary switch settings for machine and control
  2. Use of this mode for normal machining

IV. Equipment Description
  A. Main units with description of operation principle, wave shapes, voltages, important features, and required adjustments, if any
    1. Photocell reader
    2. Electronic distribution system
    3. Axis control register
    4. Relay section
      a. Axis relays
      b. Sequence relays
      c. Miscellaneous functions
      d. Feeds
      e. Speeds
      f. Tool selection
    5. Motor control
    6. Rate feedback (tachometer)
    7. Position feedback (transducer)
    8. Power supplies
      a. Lamp supply
      b. Electronic and relay supply
  B. Physical location of control elements
    1. Reader cards
    2. X-register, Y-register, Z-register cards
    3. X, Y, Z displays
    4. Lamp power supply
    5. Electronics and relay power supply
    6. Motor drive controls
    7. Motor field supply and direction switching

V. Maintenance
  A. Normal control adjustments

   1. Reader adjustments and cleaning
   2. Speed adjustments on X and Y axis
      a. Rapid
      b. Slow
      c. Creep
   3. Position-measuring adjustments
   4. Filter cleaning
   B. Fuse locations
   C. Voltage and wave-form checks
   D. Preventive-maintenance procedure

VI. Trouble Shooting
   A. General—isolation of trouble to one section of the control: It is better to think the problem out before acting. Try to have failure repeat many times to observe the area of trouble. Watch failure closely. Sometimes small discrepancies are overlooked that can narrow the area of failure.
   B. Definitions
      1. Overshoot and undershoot
      2. Readin, counting, and parity errors
      3. Manual, dial, block, and auto operation
      4. Stop on character
      5. Backlash test
   C. Use of volt-ohmmeter to measure cabinet voltages and resistances
      1. How to check diode or rectifier
      2. Check for short or open on fuses and power supply lines
      3. Check of voltages on motor drive limit (A.C.-D.C.), power supply
   D. Examples of typical troubles and demonstration of proper trouble shooting

VII. Review

The instruction is by lecture and example on a machine system, when available, or a control with a simulator, when the former is not available or convenient. A "Programming Manual" and an "Operations and Maintenance Manual" for the control are used as "texts" by the trainee. Their descriptions are quite complete so that they can be very helpful if studied and applied properly. Of particular usefulness is a chart of trouble symptoms, to be used to identify difficulties when they occur. A typical page from the trouble-shooting section in such a manual is shown in Fig. 5–11.

Item	SYMPTOM	SECTION TO BE TESTED	INSTRUCTIONS
14	Incorrect direction, feed function, spindle speed selection or tool selection being entered only	Input Unit Refer to Figure 4.2–14	Check the following: 1. The SCR Control card for proper operation of track code gates, direction code gates and SCRP pulse. 2. The SCR card containing the SCR for that particular function. 3. The particular relays associated with that function. 4. If only the direction is being entered incorrectly, check the associated track outputs of the Read Amplifier card (2A) for proper operation.
15	Read phase has ended and machining phase should begin; nothing happens	Relay Section Refer to Figure 4.2–12	Check the following: 1. MANUAL-CONTROL switch on machine panel. 2. For proper energizing of 3CR. 3. For the presence of V2. 4. For proper operation of the X and Y move relays (12CR, 13CR, 14CR, 22CR, 23CR, and 24CR). 5. 60 CR or 61CR and 70CR or 71CR for proper operation. 6. Rapid reference voltage. Refer to Section 4.2.3.1. 7. Fuse in the D.C. Motor Control Amplifiers.
16	Not Counting when axis is moved from the zero reference point.	Limit Switch Register Refer to Figure 4.2–14	Check the following: 1. Limit switch for proper operation. 2. Transducer card for proper zero head signal (HO). 3. PT Trigger card for resetting of the counter inhibit flip-flop.

Fig. 5–11. Typical sheet from trouble-shooting section of operations and maintenance manual.

A 24-hour minimum of lectures and 16 hours of laboratory work are suggested. It is also suggested that training be done at the control builder's plant or at customer schools. Training at the customer's plant is helpful but is *not* recommended as the *only* exposure to training. Experience has been that those customers who have not had training in the builder's plant have encountered the most difficulties; minor problems too readily became major crises for them.

As in every new field, a trainee cannot get too much exposure to developments in numerical control. Continued reading, studying, and exposure to equipment are strongly recommended. Many facets understood during training are forgotten after a few months away from practical applications of the knowledge. Therefore, depending upon economic considerations, refresher lectures on a periodic basis are recommended.

Following is a control manufacturer's job description for field service personnel, which, of course, is more demanding than what the typical customer requires.

JOB DESCRIPTION

Position Title: Service engineer.

Background Required: Minimum of two years, college or advanced technical school. Requires experience in field service (or development) of computer-type electronic equipment. Knowledge of vacuum tube, transistor, and relay circuits is desirable.

Primary Purpose: To do field service and maintenance of numerical controls for machine tools; position involves travel.

Duties:

1. Make periodic routine preventive-maintenance checks of controls, including cleaning of reader, taking of voltage-level readings, replacement of marginal components, etc.
2. Apply trouble-shooting steps as outlined in operations and maintenance manual, or otherwise analyze technical problems with controls in response to service call.
3. Make necessary adjustments or part replacements to put controls in good working order.
4. Make complete report, as required by company service procedure, to customer and to company service, engineering, and accounting departments. This involves obtaining signature of appropriate customer representative for confirmation of work performed.
5. Give appropriate training to customer in operation and maintenance of controls when conditions indicate preventive maintenance has not been fully practiced.
6. Give appropriate training to customer in technical facets of programming when required.

The job description must be used with care, however, because requirements vary widely with the customer. A small plant with one numerical control installation and a few personnel obviously requires a different type of individual than one of the aerospace complexes with tens of thousands of skilled personnel and many installations. It has been found that the customers who send their senior personnel for training fare best; perhaps, this is because the senior personnel are better qualified to start with, and also they probably see to it when they return to the plant that others acquire some knowledge and training.

SUMMARY

Introducing numerical control equipment involves

1. Acquiring as much familiarity as possible with the concept and its pros and cons
2. Determining whether numerical control is really needed by analyses of parts requirements
3. Specifying the technical guidelines for machine tool and control
4. Requesting quotations for specified equipment from machine-tool manufacturers
5. Establishing the economics of justification
6. Preparing for prospective purchase by establishing maintenance and service guidelines

Interspersed among the foregoing should be appropriate dissemination of information to interested personnel so that acquisition of numerical control equipment does not come as a surprise and there is an opportunity to appreciate its contribution to the company's and employees' well-being, growth, and security. The procedure here varies widely, depending upon the size and type of company. It is well to acknowledge possible dislocations and to prepare for them in order to avoid unnecessary loss or inconvenience to company or personnel. Finally, all numerical control equipment to be acquired should be looked at with an eye toward its ultimate conformity to an overall advanced manufacturing system.

Unquestionably, misapplication of, or lack of adequate preparation for numerical control can be troublesome and costly. On the other hand, awareness of the basic concepts involved and an orderly approach in evaluating the alternatives can yield large rewards not feasible by any other means.

6

Machining Centers and Integrated Systems

The evolution of numerical control to its present state of universal acceptance in manufacturing has taken place in a relatively short period of time. It has come a long way from the laboratory curiosity of a little over a decade ago. Yet, in a sense, only the surface has been scratched toward achieving its full potential. Daily, new ground is being broken toward this end. It is useful to review some of the trends reflected by the efforts currently being applied in the design, implementation, and use of new numerical control equipment.

Machine Tools

The magnitude and breadth of numerical control's impact in manufacturing are exemplified by the virtual revolution it has prompted in the design of machine tools—an industry considered ultraconservative and highly resistant to change. The nature of automatic control has made demands on design for greater precision, greater rigidity, part holding and index-ing, and lubrication. Most important has been the need for new drive concepts (to be amenable to control signals). Elimination of operator manipulation to cause machine movements obviates the need for handwheels (at least permanent ones) and levers. With manual machines, these had to be conveniently located within the operator's reach; therefore, they greatly influenced overall machine configurations, which stood substantially unchanged for many years. With numerical control, desirable machine concepts previously considered impossible have become practical.

What are the continuing trends? Certainly, greater precision and rigidity are both desirable. The lack of operator intervention with numerical control means that precision must be an inherent characteristic of the machine. Accuracy is not limited by operator response capability (and seldom by control capability). Consequently, greater machine precision is transferable to machining precision with numerical control, so it will continue to be sought.

Machine vibrations and tool chatter take on special importance with numerical control because of the greater range of speeds and feeds used; that is, numerical control is used for more extreme conditions of machine and cutter loads. This, of course, is highly desirable, but it is also more demanding of rigidity. More rigidity results in greater machining capability via the handling of tougher materials and higher cutting rates.

Means for part handling in general are being incorporated in machines and machine systems. These include pallets, shuttles, and index tables with clamping devices for part holding, feed devices, part-orientation means (indexing and continuous), and transfer mechanisms. In fact, numerical control allows essentially free choice of parts and selection of operations to be performed. It makes feasible sophisticated multiple transfer systems with part, operation, and machine codings to facilitate alterations in sequences and routings.

Machine designs are exhibiting an increasing tendency toward building-block construction, with more thought being given to interchangeability of the blocks. The controls of numerical control do not differentiate among various machine types; that is, the control parameters for many machines are similar enough that much interchangeability can be exploited with numerical control as long as the corresponding machine elements are designed to have similar input and output interfaces. There are machine tools, for example, with movement requirements along three orthogonal axes, that are quite similar. The fundamental difference among these tools is in the cutters. Consequently, a change of a head or spindle accommodating a cutter amounts to a change of machine.

Even more dramatic is the tendency to design machines with automatic means for changing cutters while in operation. These are machines that incorporate "tool changers." Since the capacity and variety of tools (cutters) are large, the new machines are quite general in functional capability. The flexibility made possible by numerical control makes these machines practically complete shops in that they are often capable of performing all operations required on a part without need for other machines. Accordingly, these are known as "machining centers."

Machining centers represent a revolutionary development in the design of machine tools. In just a few short years, extremely versatile machining centers have been introduced. The trend is definitely toward more of them, from very simple relatively inexpensive versions, to allow small shops to avail themselves of them, to quite elaborate, expensive models. This is in keeping with the tendency in manufacturing to require men and machines that are generalists as opposed to specialists.

The list of machines that might be called machining centers is quite long, depending upon the liberties taken in qualifying the automatic tool changer. Strictly speaking, the latter can include merely straightforward indexing means for presenting one of a variety of tools in response to an external command even if the command is simple sequence control, power range is limited, tool types are similar, and tool capacity is as low as two. Yet, some tool changers are very sophisticated in providing quite random selection from a large variety of tools capable of doing significantly different cutting jobs over a broad power range. Most generally, machining centers are used to perform automatic operations on parts under tape control, including milling, drilling, tapping, reaming, and boring. Machining is done on the parts in place, with tool changes also accomplished automatically under tape control.

Following are some of the critical specifications and comments on prominent representative machining centers:

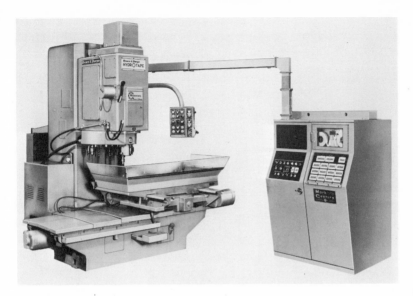

Fig. 6–1. Brown & Sharpe Hydrotape machining center. (Courtesy of Brown & Sharpe Manufacturing Company.)

Brown & Sharpe Hydrotape 232, 233, and 234. The Hydrotape machining centers (Fig. 6–1) are available with automatic tool changers but also in versions for manual tool changing. On Hydrotape machines, the carrier indexes and changes tools in approximately 3 seconds. Only ½ second is required to skip an unused station. The tool holders are automatically inserted and held in position in the spindle nose by hydraulic pressure. Tools are quickly and easily loaded into the carrier track at the gate on the side of the positioner.

Specifications

Table size	Up to 57½" × 26"
Longitudinal (X) travel	20", 30", and 48"
Cross (Y) travel	15" and 25"
Vertical (Z) travel	13"
Stroke of quill	12½"
Tool storage	12
Maximum tool size	Up to 1⅞" dia. drill
Spindle speeds	24, from 65 to 3000 rpm
Feeds	79, from 0.5 to 39.5 ipm

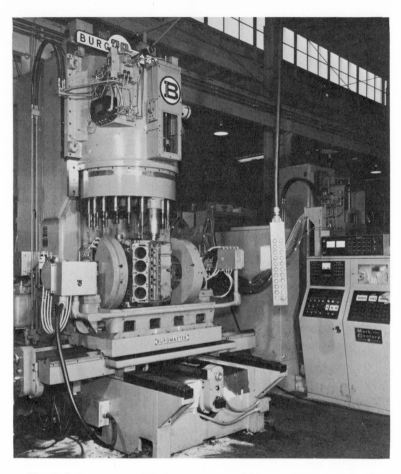

Fig. 6–2. Burgmaster 20T-SH. (Courtesy of Burgmaster Corporation.)

Burgmaster 20T-SH. The Burgmaster 20T-SH (Fig. 6–2) is a sliding-head, master-quill type of machining center. Random tool selection of 20 tools is available, with tape selection automatically providing for the shortest distance traveled. Another type is a traveling-column version (Fig. 6–3), which, teamed up with a heavy-duty indexer, allows work on four sides of large parts. The traveling column permits load and unload of heavy components during the machining cycle.

Fig. 6–3. Burgmaster 20T-SH-TC. (Courtesy of Burgmaster Corporation.)

Specifications

Table size	31″ × 46″
Longitudinal (X) travel	45″
Cross (Y) travel	30″
Vertical (Z) travel	18″
Quill stroke	12″
Tool storage	20
Maximum tool size	5″ (can be up to 11″ by leaving out tool in adjacent spindle)
Spindle speeds	24 speeds 99 spindle feed rates

Cincinnati 24 ATC. In this machine, the column moves sideways on the bed (X axis), and the cutter spindle slides up and down on the column ways (Y axis). The index table has eight positions.

Fig. 6–4. Cincinnati 24 ATC. (Courtesy of Goodyear Aerospace Corporation.)

The tool-storage drum is on the rear of the column and is stationary. A preload arm transfers the tool from the drum into the tool changer, which also operates on the stationary column

Specifications

Index table	24″ × 24″
Longitudinal (X) travel	32″
Cross (Z) travel	26″
Vertical (Y) travel	26″
Tool storage	35
Maximum tool size	3″ dia. × 14″
Position accuracy	0.0005″
Repeatability	0.0003″
Spindle speeds	16, from 33 to 2000 rpm
Feeds	0.5–100 ipm

Fig. 6–5. Cincinnati 24 ATC. (Courtesy of Goodyear Aerospace Corporation.)

(Figs. 6–4 and 6–5). The spindle always returns to the "home" position for the tool change. The tool changer provides automatic selection of up to 99,000 different tools. This has the advantage of allowing some of the repeat tools to be left in the machine. This also permits cataloging and storing by the code number, facilitating programming of tools by assigned numbers (Fig. 6–6).

Cleereman Spindlemaster. The tool magazine is a disk equipped with pockets, which rotates on a centerpost to present any one of 30 tools to the machine spindle. Each tool pocket around the perimeter of the carrier disk is numbered, from 1 through 30. A tool is selected in any sequence by the pocket number in which it is placed. When a tool is called for, the magazine rotates either clockwise or counterclockwise (Fig. 6–7).

Fig. 6–6. Cincinnati 24 ATC. (Courtesy of Goodyear Aerospace Corporation.)

Fig. 6-7. Cleereman Spindlemaster SP-24. (Courtesy of Cleereman Machine Tool Corporation.)

167

Specifications

	Model SP-24	Model SP-42
Table size	16″ × 30″	16″ × 56″
Longitudinal		
(X) travel	24″	42″
Cross (Y) travel	16″	16″
Vertical (Z) travel	9″	9″
Spindle travel		
(vertical)	9.9999″	9.9999″
Tool storage	30	30
Maximum tool		
size	4″ dia.	4″ dia.
Position Accuracy	±0.001″	±0.001″
Repeatability	±0.0005″	±0.0005″
Spindle speeds	79, from 40 to 3200 rpm	79, from 40 to 3200 rpm
Feeds	79, from 0.5 to 39.5 ipm	79, from 0.5 to 39.5 ipm

Ex-Cell-O Work Center. The machine is basically a horizontal spindle machine. The workpiece is fastened to a T-slotted table top which can be indexed in 1° increments (B axis). The table moves in and out toward the spindle (Z azis) on a saddle assembly. This saddle assembly moves along the longitudinal X axis on a rigid machine bed. Attached to the bed is a rear column to and from which the cutter-spindle slides vertically (Y axis). The tool-changer assembly is an integral part of the cutter-spindle assembly and also slides vertically (Fig. 6–8). The tool storage is an oblong arrangement permanently affixed to the side of the column. It is capable of storing 32 tools.

In order to accommodate the removal of the tools from the tool storage and their placement in position in the upper right-hand corner of the machine column and at right angles to the tool storage, there is an intermediate "tool turn-around assembly" (Fig. 6–9). In order to execute a tool change, the cutter-spindle assembly returns upward to its home position, at which point the interchange is executed in 5 seconds because the "next" tool is in a "ready" position. This arrangement places the cutter, workpiece, "next" tool, upper part of the tool storage, pendant station, and control unit within easy full view

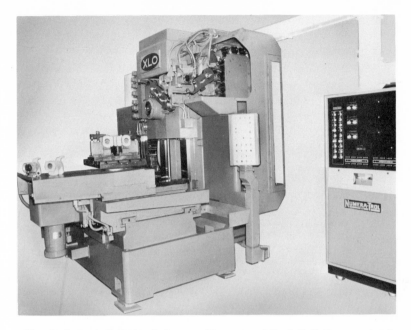

Fig. 6–8. Ex-Cell-O Work Center. (Courtesy of Ex-Cell-O Corporation.)

of the operator, without necessitating any additional movements on his part to relocate himself physically.

Specifications

Longitudinal (X) travel	28″
Vertical (Y) travel	20″
Cross (Z) travel	18″
Table (B)	20″ × 20″
Tool storage	32
Nominal maximum tool size	5″ diameter
Interchange time	5 seconds
Resolution	.0001″
Repeatability	±.0002″
Net accuracy	±.0005″
Spindle speeds	85–2040 rpm, infinitely variable
Feed rates, linear axes	0–50 ipm
Rapid-traverse rate	200 ipm
Rotary table	1 revolution/10 seconds

Fig. 6–9. Ex-Cell-O Work Center. (Courtesy of Ex-Cell-O Corporation.)

The cutter-spindle slide is supported by two hardened and ground rails, which face the workpiece; thus, the thrusts from metal cutting are supported directly into these two rails and induce minimum torsion to the column. The feeds are from ½ to 50 ipm at ½-inch increments.

Giddings & Lewis NumeriCenter 15-V. This machining center was developed specifically for automatic short-run work. Its

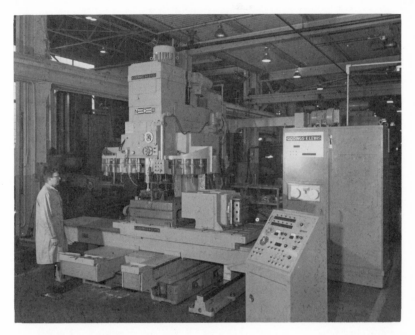

Fig. 6–10. Giddings & Lewis NumeriCenter 15-V. (Courtesy of Giddings & Lewis Machine Tool Company.)

key feature is the dual-matrix automatic tool changer, providing capacity for 40 tools (Fig. 6–10).

The drilling cycle on this machine automatically advances the spindle at 125 ipm until it contacts the workpiece; the feed automatically engages; the spindle drills to depth and retracts at 200 ipm. With this system, axial presetting of tools is not re-

Specifications

Table size	Standard: 38″ × 78″
	38″ × 90″
Index table	Rotary indexing fixture optional
Tool storage	40: 20 each in dual matrix
Maximum tool size	Body dia. up to 5″
Spindle speeds	2 ranges, from 15 to 3000 rpm
Feeds	3 ranges, from 0.001″ to 0.158″
	per revolution

quired, because depth measurement does not start until the tool contacts the workpiece. The machine is also equipped with a control that provides the same basic advantage in tapping operations without screw or geared thread leads.

Giddings & Lewis NumeriCenter 25-H. The automatic tool changer travels with the head, thereby permitting tool change to occur anywhere within the travel limits (Fig. 6–11). This feature enables the head to stay locked on the center line of the hole throughout sequential machining operations on that hole, eliminating repeatability errors in repositioning. The large tool storage is a key feature—even 8-inch-diameter cutters can be stored by leaving an adjacent tool chamber empty.

Also available is automatic radial orientation of each tool every time it is replaced in the spindle. This eliminates variations in bore size due to runout in the tool that can occur when tools are changed.

Fig. 6–11. Giddings & Lewis NumeriCenter 25-H. (Courtesy of Giddings & Lewis Machine Tool Company.)

Specifications

Cross (X) travel	60"—up to 132"
Longitudinal (Z) travel	48"—up to 94"
Vertical (Y) travel	48"—up to 72"
Tool storage	63, in 2 magazines: 32 outer
	31 inner
Maximum tool size	4½" dia. × 18" in inner
	6" dia. × 18" in outer
	Tools weighing up to 75 lb
Spindle speeds	32, with full hp across total range
Feeds	36, in ipm to saddle, table, and headstock

Hughes Aircraft Company MT-3. The 18-inch-diameter work-positioning table may be rotationally indexed to 16 positions either clockwise or counterclockwise at the programmer's option (Fig. 6–12). The MT-3 is arranged so that up to three different spindle head systems surround the workpiece, permitting the work to be "played in and out of the head assemblies" so that, while one head is cutting, the next head to be used can be made ready.

Specifications

Index table	18" dia.
Cross (X) travel	24"
Longitudinal (Z) travel	19"
Vertical (Y) travel	12"
Tool storage	30 in universal head
Maximum tool size	3½" dia.
	Larger diameters can be changed automatically by leaving adjacent pockets empty.
Position accuracy	±0.001"
Repeatability	±0.0003"
Spindle speeds	28, from 40 to 4000 rpm
Feeds	31, from 1 to 75 ipm
Work size	12" × 24" × 24"

Optional machine-tool "building blocks" are a heavy-duty milling head, a precision-boring head with tool changer (Fig.

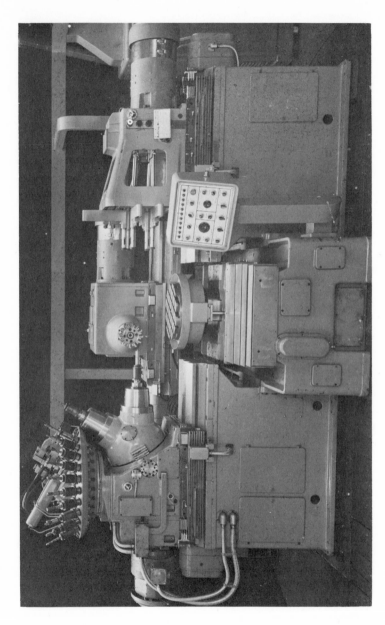

Fig. 6–12. Hughes Aircraft Company MT-3. (Courtesy of Hughes Aircraft Company.)

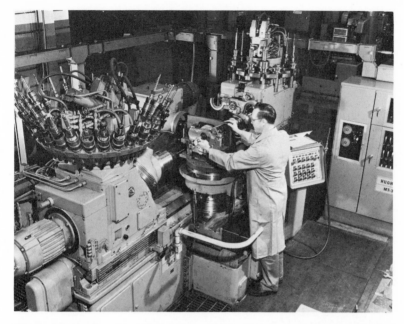

Fig. 6–13. Hughes Aircraft Company MT-3. (Courtesy of Gleason Works.)

6–13), and a multiple-spindle drilling head. The "building block" design makes field addition of other elements practical.

Precision-Boring-Head Specifications

Travel	19"
Tool storage	14
Maximum tool size	6" dia.
	Larger diameters can be changed automatically by leaving adjacent pockets empty.
Spindle speeds	21, from 80 to 2400 rpm
Feeds	31, from 0.5 to 37.5 ipm
Diametral accuracy	0.0005" TIR on 1" bore

Kearney & Trecker Milwaukee-Matic II. The cutter head is a compound slide arrangement, moving up and down on the Y axis. Travel in the Z axis is by moving in and out of the horizon-

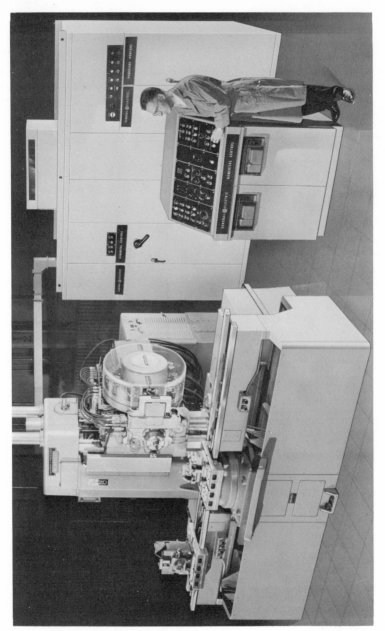

Fig. 6–14. Kearney & Trecker Milwaukee-Matic II. (Courtesy of Kearney & Trecker Corporation.)

tal portion of the slide carrying the spindle. The tool changer is part of this horizontal slide (Fig. 6–14). The table indexes to 8 positions or 360,000, depending upon the model. There are positioning and contouring versions, with one model equipped for five axes of control. There are also larger versions, which are known as Milwaukee-Matic III and Milwaukee-Matic V. One model of the former, for instance, has controlled motions of 44 by 44 by 50 inches, has a 42-inch-diameter rotary table, can accommodate tools up to 6 inches in diameter, and has five axes of control, including table tilt to 90,000 positions.

Specifications

Index table	18″ × 18″ pallet
Longitudinal (X) travel	24″
Cross (Z) travel	16″
Vertical (Y) travel	20″
Tool storage	31
Maximum tool size	2⅝″ dia. × 7½″
Position accuracy	±0.0005″
Repeatability	±0.0002″
Spindle speeds	100–4000 rpm, in 10-rpm increments
Feeds	0.5–9.9 ipm, in 0.1-ipm increments
	10–50 ipm, in 1.0-ipm increments

Kearney & Trecker Milwaukee-Matic Series E. Figure 6–15 pictures a Milwaukee-Matic Series E. It is also available with tool changer and a one-tool manually loaded ready position, instead of the tool-storage magazine. Both models can be equipped with a plain table or with a four-position index table.

Specifications

Table size	16″ × 49″
Index table	14″ × 14″
Longitudinal (X) travel	24″
Cross (Z) travel	14″
Vertical (Y) travel	14″

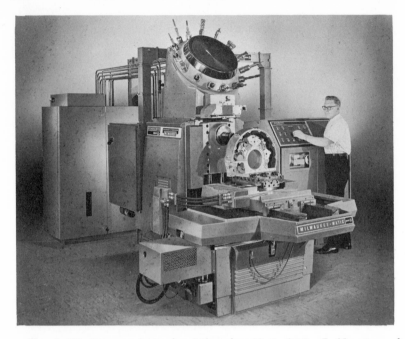

Fig. 6–15. Kearney & Trecker Milwaukee-Matic Series E. (Courtesy of Kearney & Trecker Corporation.)

Tool storage	15
Maximum tool size	4″ dia. in 7½″
Position accuracy	±0.001″
Spindle speeds	32, from 100 to 3000 rpm
Feeds	1 to 9.9 ipm, in 0.1-ipm increments
	10 to 70 ipm, in 1.0-ipm increments

Pratt & Whitney MC 1000. To carry the biggest loads in boring operations, the X axis is the carriage with the longest rails (Figs. 6–16 and 6–17). The maximum work cube is 16 inches square with unlimited height. Table index accuracy is ±0.10 second of arc. In addition to the indicated feeds, optional contouring feeds of from 1 to 50 ipm are available. Numerical control in three axes for either point-to-point or contouring is available in addition to the facility for totally manual machining.

Fig. 6–16. Pratt & Whitney MC 1000. (Courtesy of Pratt & Whitney Machine Div. of Colt Industries.)

Specifications

Table size	16″ × 24″
Index table	16″ × 16″
Longitudinal (X) travel	24″
Cross (Z) travel	16″
Vertical (Y) travel	16″
Tool storage	15
Maximum tool size	4″ dia.
Position accuracy	±0.001″
Repeatability	±0.0005″
Spindle speeds	30–3000 rpm (infinitely variable)
Feeds	1–70 ipm

Sundstrand Omnimil OM-2. The 54-inch-stroke linear table (X axis) with T slots carries a built-in, flush-top rotary index table (Fig. 6–18). Long parts can be mounted across the entire length of the linear table. Tape-selected indexing (eight posi-

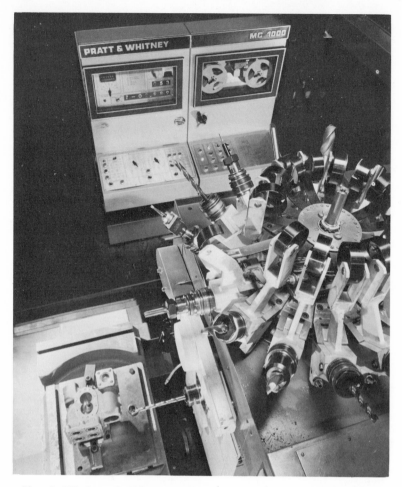

Fig. 6–17. Pratt & Whitney MC 1000. (Courtesy of Pratt & Whitney Machine Div. of Colt Industries.)

tions) is provided for the standard rotary table. As an option, numerical control of the rotary table, providing 1 million indexing positions and feed for point-to-point or contour machining, can be furnished.

The automatic tool changer accommodates 39 tools up to 2¾ inches in diameter—or a smaller number of larger tools up to 6 inches in diameter. Five-digit key coding permits use of 99,999 different tool numbers.

Fig. 6–18. Sundstrand Omnimil OM-2. (Courtesy of Sundstrand Corporation.)

Several makes of numerical control systems are available, utilizing punched paper tape for both point-to-point and contour machining.

Specifications

Table size	24½″ × 68″
Index table	24″ dia.
Longitudinal (X) travel	54″
Cross (Z) travel	24½″
Vertical (Y) travel	27″
Tool storage	39
Maximum tool size	2¾″ dia. × 18″
Position accuracy	±0.0005″
Repeatability	±0.00015″
Spindle speeds	10–110 rpm (1–10 hp)
	110–2000 rpm (10 hp) (15 hp optional)
Feeds	
Contouring	0–2 rpm
Positioning	2 rpm

Sundstrand Omnimil OM-3 and OM-4. These five-axis machining centers (Fig. 6–19) are capable of performing more

Fig. 6—19. Sundstrand Omnimil OM-3. (Courtesy of Sundstrand Corporation.)

Fig. 6–20. Sundstrand Model 21 Jigmatic. (Courtesy of Sundstrand Corporation.)

complex work on larger parts than the OM-2. They can machine any contour at almost any angle. Sundstrand also makes a three-axis rail-type machine with a 20-position automatic tool changer, named the Model 21 Jigmatic (Fig. 6–20). A 7½- or 10-horsepower machine, it can mill, drill, bore, ream, or tap.

Specifications

	Model OM-3	Model OM-4
Rotary table	42″ dia.	52″ dia.
Longitudinal (X) travel	48″	96″
Cross (Z) travel	48″	72″
Vertical (Y) travel	48″	96″
A-axis travel	150°	150° or 360°
C-axis travel	360°	360°
Tool storage	59	59
Maximum tool size	11″ dia. × 24″	11″ dia. × 24″ or larger
Position accuracy	±0.0005″	±0.0005″
Repeatability	±0.00015″	±0.00015″
Spindle speeds	Low: 12–1000 rpm	
	High: 50–4000 rpm	
Feeds	.0 to 200 ipm	

Sensors and Controls

The need for greater precision and reliability means that better sensors and measuring techniques are required. Not only is greater accuracy desirable, but the means for measuring must be amenable to use in automatic systems by yielding output signals usable by controls. Work is proceeding along many lines in developing new methods, including expanded usage of laser technology. Non-contact techniques (mainly optical) are increasingly being used in replacing conventional mechanical means of measuring and gaging.

More of the gaging is being done "on line." Traditional methods of checking dimensions after parts have been completed and unloaded are being replaced in some cases by means for gaging parts while they are in process. The gaging data are usable in modifying the process itself to assure conformity to dimensional specifications.

There is growing interest in monitoring more of the parameters in the cutting process so that the data may be used as feedback in controlling the process. This is generating research and development in many areas including the metal-cutting process itself; the sensing devices for measuring forces, temperatures, and dynamic characteristics involved in various cutting situations; and the control logic so that it may utilize the sensed data appropriately.

Along control lines, considerable effort is being expended in ultraminiaturization through the application of microelectronics. The technology is known as "microelectronics" because it involves the reduction of circuit components to such an extent that they are essentially invisible to the naked eye; that is, the dimensions are microscopic. The technology is also referred to as "integrated circuitry" because of the inseparable association of the circuit elements.

One of the reasons for activity in miniaturization is, of course, to yield a smaller overall control package. However, even more important reasons are tied to the fact that numerical control equipment innovation follows computer art where integrated circuits contribute to ruggedness, reliability, economy, and performance of the systems into which they are incorporated. They make it possible to construct control equipment modularly where standard modules are available. Yet the modules represent larger functional pieces than do conventional components. Microcircuitry will continue to be involved in newer compact and high-speed controls.

A comparatively new field, fluidics, is receiving increasing attention for certain implementations of numerical control equipment. Fluidics involves the control of direction and flow in a continuously moving fluid by injection of a small amount of fluid into the main stream. The fluid is most often air, although it can be water or other liquids or gases. The injected fluid need not be identical with the main stream fluid. Fluid amplifiers are devices with variously configured passages, which allow functions to be performed as the fluid passes through them. The devices can be assembled in a variety of ways to form logic networks much like those in electronics. Consequently, electronic functions can be duplicated in some respects with fluidic devices.

Interest in fluidics is based on the insensitivity of fluid devices to corrosive, high-temperature, and radioactive environments, which might affect electronic equipment. No moving parts are involved so that breakdown is not a problem. The devices are simple, reliable, and inexpensive. The biggest problem encountered with fluidics is that response speeds are significantly lower than those available with electronics. Therefore, high-speed applications are relatively limited. On the other hand, the simplicity and economy are tangible enough to generate much continued effort in new systems and applications based on fluidics.

MAN-MACHINE RELATIONSHIPS

A concept that will be increasingly involved in the design of machines of all types is the so-called man-machine relationship. This has to do with considerations of mutual effects between humans and machines. For numerical control systems, there are a number of places where the efforts of man and machine converge. We have already discussed machine outputs and their effects on the operator, the manufacturing department, and other elements of a company's organization. We have mentioned too the interface at machine input where dials, push buttons, etc., may provide the data used by the machine system. The main input is punched tape, which represents the communication link between man and the machine system. The language of the punched tape is such that it can be interpreted by the control of the numerical control system to direct the machine tool through its motions and functions.

Computer-assisted Programming

As pointed out before, input to the punched tape used in numerical control is part-programming information prepared by the planner or programmer. There are times when the data to be programmed comprise so many detailed points that programming in the classical manner is virtually impossible. Manual computations may represent an enormous amount of work, which makes the program very tedious to prepare and easily susceptible to error. Situations of this type occur in both contouring and positioning work.

In the foregoing cases, the part program is most often amenable to being broken down into fundamental operations in terms of certain standard routines, which themselves are amenable to computer treatment for carrying out repetitive subroutines. The part program can then be put into an abbreviated form, if the details are carried out by a computer. Therefore, a general-purpose computer program (or compiler) can be prepared for carrying out the details assumed by the abbreviated program for the part. For example, the computer program may contain instructions for a general-purpose computer to calculate points in the desired cutter path. Finally, since the computer program is for general computations without reference to the specific control to be used, additional means are required for relating the part program and the computer program to each specific control. A program to accomplish this correlation is known as a *postprocessor*. The part program, the computer program, and the postprocessor are normally used as inputs to a general-purpose computer whose output is a tape with a format directly usable by the numerical control system. If the computer output is magnetic tape, an additional step may be required for conversion to punched tape usable by the control.

Many computer programs are available—some with corresponding postprocessors, others without them. There will continue to be activity in this area. Perhaps the biggest problem to be resolved is that the many programs available have been generated with so wide a variety of computer equipment in mind. Universally accepted standards for computer programs and postprocessors are difficult to achieve because of the lack of availability of every type of computer at all locations.

Adaptive Control

A rather new concept applied to numerical control is adaptive control. A truly adaptive system is one that monitors the environment of the machine system and causes the system to correct itself if undesirable changes occur. With conventional machines, the operator provides adaptability to the system by changing machine conditions when any of the machining parameters exhibit characteristics contrary to what he believes to be good practice. The operator may observe chatter and proceed to slow down the feed to decrease cutter load, which may have prompted the chatter. He may determine that workpiece surface characteristics are not as they should be, so that temperatures at the cutter-workpiece interface are too high. Therefore, he may decide to use a different spindle speed for better cutting. The operator provides adaptability any time he observes a situation, decides from his experience that the situation requires a change of machine conditions, and takes corrective action.

Since numerical control systems do not ordinarily rely on operator intervention, adaptability in accordance with the foregoing must be provided within the numerical control system itself. Sensors can, indeed, correspond to the operator's observations, and control logic can be designed to accommodate corrections as required. The dictation of corrections in accordance with an evaluation of the observations compared to past experience is, however, difficult to implement automatically. This is not to say that it cannot be done. In fact, considerable research is being done on self-learning systems, which memorize past observations so that repetitions have greater influence on the types of corrections to be made. However, these are not yet sufficiently well developed to be practical.

The type of adaptive control applied to numerical control is a simplification of the foregoing. Certain machine parameters are sensed, automatic comparisons are made with programmed allowable levels for these parameters, and corrections to the machine system are made in a preordained manner if these levels are exceeded. A simple example of this is a system which would provide for automatic changes of spindle speed with changes in automatically measured torque at the spindle (which assumes the latter represents the machining criterion of interest).

Several adaptive systems are in use, but each is quite simple and concerned with only one or two machining parameters. Eventually, effective adaptability must be based on many more parameters. The difficulties to be resolved are

1. There are many parameters (perhaps 50 or more) that may influence machinability
2. The machining system at the cutter-workpiece interface is complex so that machining conditions used, such as feeds, speeds, tool material and configuration, and coolant, are still empirical. This means corrective action involves a high degree of trial and error.
3. The sensing means for some of the machining parameters (e.g., tool-chip interface temperature) require considerable development before they can be relied upon in a production sense.

In any event, much effort is being devoted to adaptive control, and at least partial adaptability, though quite minor in some cases, will be incorporated in the numerical control systems.

Numerical Control—Computer Interconnections

There is no question that the future will witness a more complete marriage between computer control and numerical control.[1] The idea of the entire manufacturing process' being

[1] Julian E. Wilburn, "Future Marriage of N/C and Computer Control," with comments by William C. Leone, *Automation*, January, 1966, pp. 79–83.

handled as a system, rather than as a collection of separate quasi-independent entities, is a certainty. This will eventually allow numerical control to be treated along with production-control aspects (such as scheduling, work-load, determination, and assembly) and accounting in an automatic manner, and thus plant control can be achieved. The computer can correlate the myriad details to optimize the process, fully report the results, and call for any action required of management.

We must not lose sight of the fact that many obstacles still exist to prevent reaching this goal for some time yet. In the first place, the manufacturing process is somewhat difficult to define in many industries. Secondly, the trend in manufacturing has been, and will undoubtedly continue to be, toward small lots of many varieties. Although this fact emphasizes the need for the marriage between numerical control and computers, it also imposes an immediate requirement for flexibility of approach, which makes implementation difficult.

Then, too, we must recognize that planning for maximum exploitation of control tools has not been as rapid as it might be. Although good equipment has been available, we have been relatively slow in designing for numerical control. Finally, even after the marriage is performed, there will be considerable practice of the "double standard." There will always be many shops, both large and small, where the inherent nature of the operations will preclude a complete process-control approach. Getting into and out of the control system sometimes places too heavy a burden on the simpler, straightforward jobs.

In any event, greater participation of the computer in broadening the impact of numerical control is certainly the trend. For instance, work is already being done in expressing part descriptions in terms of definitions and dimensions, as opposed to putting them in drawing form. The computer can communicate in this language and, therefore, obviate the need for many of the steps presently required in preparing instructions for the controls of numerical control.

PLANNING FOR THE FUTURE

Certainly the ramifications of the impact of numerical control on manufacturing and other plant practices are many. What can management do now to prepare for the future? What are the guidelines? Simply, management must educate itself, keep itself up to date, establish a long-range plan, and set up communications programs for the worlds both inside and outside the plant.

Automation does not limit its influence to the plant itself. Rather, it has broad social consequences. Therefore, management bears a responsibility to society to harness its favorable potentials and to forestall any possible detrimental aspects. Management cannot do this unless it keeps itself informed about the newer concepts and corresponding equipment that will be available. It is no longer adequate to leave equipment selection to the shop foreman as if his shop were a separate domain independent of influences on other departments. Decisions on the tools of production are top-level responsibility now. Therefore, management must be as concerned with the "what" and "how" of these tools as it has been with merely the "how much" in the past.

A long-range plan for automation is an absolute necessity. Such a plan recognizes that numerical control must ultimately fit into an overall integrated manufacturing system. Acquisitions of numerical control equipment should not be considered piecemeal. Just as every well-managed company establishes long-range targets for product areas and markets it wishes to serve, it must establish a philosophy for acquiring the tools it will require to achieve those goals. It must design a system to outline the long-range capital investment it will make to suit its established acquisition philosophy. Technological advancements and the dynamic nature of the world economy preclude the design of a plan that cannot tolerate some modifications as conditions change. Therefore, management's long-range plan must be kept current at all times.

Purchasing new equipment is not enough. Management must generate a communications hierarchy within the company to disseminate appropriate information to employees involved directly or indirectly with changes in equipment or practices. Training in the proper use of new equipment and the concomitant effects on procedures must be adequately arranged for sufficiently in advance of its introduction to the plant to avoid serious misapplications. The employee-relations aspects of procedural and organizational changes must also be thoroughly studied prior to the changes' being put into effect.

Management must also carry out its responsibility to participate in the promotion of public understanding of such automation concepts as numerical control. Cooperative government and labor communities will enhance their potential. There must be a more universal appreciation that numerical control is not mainly intended to eliminate labor, but that the greater productivity and flexibility of numerical control result in more and better-quality products available to more people at less cost— vital ingredients for a higher standard of living.

7

Representative Numerical Control Applications

MACHINING LARGE WORKPIECES

MACHINE: Three-axis boring mill
CONTROL: Positioning plus displays

For large parts, numerical control machine performance is not measured in "parts per hour." The size of the workpieces and the sequence of operations performed on them preclude any measurement except "work per hour." The workpieces in the floor-type three-axis boring mill shown in Fig. 7–1 are machine-tool castings—bases and columns—that are subsequently assembled into precision gear-forming machines. The machine, in this sequence, mills mounting surfaces and way beds using 8- and 12-inch-face milling cutters; it cuts keyways and reliefs with end mills; then it spot-drills in more than 60 locations in ten rows, and drills and taps holes of various diameters at each of the spots.

The entire sequence of operations is programmed in one tape, including the changes in feeds and speeds of the column headstock and the spindle, and retraction and reverse rotation of the spindle to withdraw the tape.

From the control station on the vertical column of the machine, the operator positions one of two worktables. Then he turns the job of machining the work over to a numerical po-

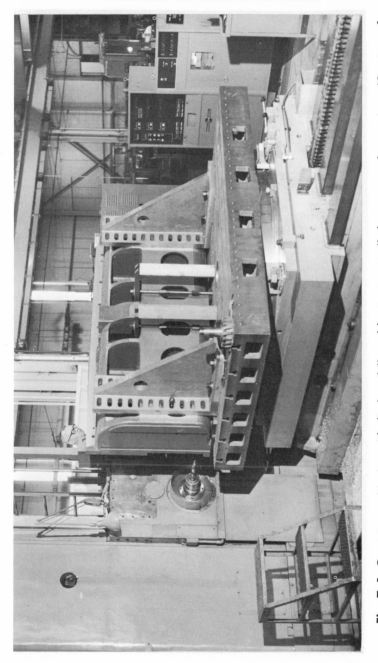

Fig. 7–1. Operator positioning with displays followed by tape-controlled sequence of operations. (Courtesy of Michigan Tool Company.)

sitioning control. Besides the control, the system is equipped with a position-display device on which the operator relies for accurate positioning of the worktables before assigning the machining responsibility to the control.

While the mill is machining the work on one table, the second table is loaded and positioned. The position display measures the movement of both tables in individual banks of numerals providing precise "line up" capabilities.

TUBE BENDING WITH NUMERICAL CONTROL

MACHINE: Tube bender
CONTROL: Positioning

Improved accuracy, additional versatility, and automatic operation resulting in decreased handling and setup times are a few of the advantages accruing from a combination of tube bender and numerical positioning control.

Simplified setup is realized by the control in automatically returning to known settings after each operation. Every machine operation, including reference returns, is digitally programmed in the numerical control system.

The numerical control bending machine is programmed to take a straight tube from the operator, insert a mandrel, and make bends in a sequence, while automatically positioning for plane of bend and distance between bends (Fig. 7–2).

As in most bending processes, springback was the prime difficulty to be overcome. This problem is solved by multiplying the bend angle read from paper tape by a proportional factor which is dependent upon the bend angle and the material being bent. The composite bend angle is further added to another constant factor, which is a function only of the material being bent. This means that one punched tape with the desired angles can process pipe of different sizes and various materials, all requiring the same bending, by merely applying these correction factors.

Recent bending-press advances and the adoption of numerical control have enabled fabricators to standardize and tighten bend radii. The adoption of a single bend radius for a single part reduces tooling inventory and makes progressive high-speed

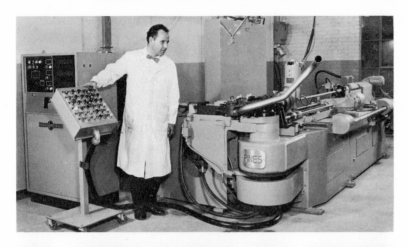

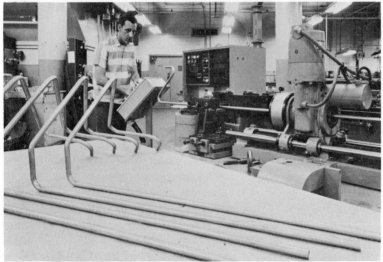

Fig. 7–2. Tube bending with numerical control. (Courtesy of Pines Engineering Company.)

bending possible. In addition to increasing speed, the numerical control bending press can reduce the warehousing of semifinished materials stock.

Sharper bends and more intricate configurations and variations in bend planes are now demanded in order to make more salable and functional shapes. As these radical bends and new

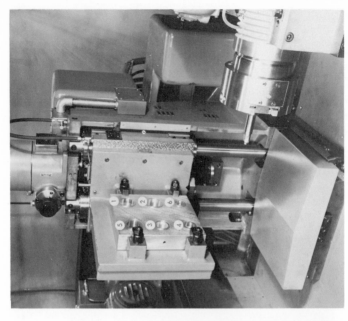

Fig. 7–3. Double-end, single-point boring: *left*—double-end boring tool with test part in place; *right*—close-up showing fixture and test part mounted on vertical slide. (Courtesy of Ex-Cell-O Corporation.)

materials are called for and as the bending technology encompasses new shapes, it is found essential to incorporate numerical control. The system can meet these greater performance requirements while performing at reduced cost.

DOUBLE-END, SINGLE-POINT BORING

MACHINE: Double-end single-point boring tool
CONTROL: Four-axis positioning

One system performs single-point boring operations from either end with variable bore diameters within the range of the machine (Fig. 7–3). Straight and step boring and counterboring operations, accomplished from either end of the machine, are tape-programmed to bore progressively from the largest to the smallest diameter.

A test part with six holes is used to illustrate the machining capabilities and to suggest the full potentialities of this type of application. Hole locations on the test part have been numbered to facilitate following the sequence of operations.

The tape-controlled machining cycle is programmed to bore holes 1, 2, 3, and 4 in sequence from the left-hand end and holes 1, 5, 4, and 6 in sequence from the right-hand end of the machine. Holes 1 and 4, counterbored from both ends, demonstrate the machine's ability to hold concentricity within extremely close tolerances. The ability to change diameter by tape controlled programming is demonstrated by step boring two different size diameters during operations on holes 5 and 6. With conventional equipment, the accuracy and repeatability demanded is almost impossible to achieve because of repetitive part locating requirements.

MACHINING LONG, NARROW PIECES

MACHINE: Open-side, traveling-head turret drill
CONTROL: Two-axis positioning

For channel bars which are 12 feet long and 3 inches wide, it was previously necessary to use six different table-type, single-spindle drill presses. With a tape-controlled, open-side, traveling-head turret drill, it is possible to set three channel bars on the table and make three 12-foot passes of the turret head, left to right, right to left, and left to right (Fig. 7–4).

Fig. 7-4. Traveling-head turret drill; view at right shows channel bars in place. (Courtesy of Automatic Electric Company.)

In addition to increasing production, the numerical control system eliminated excessive handling of parts, reduced the floor space required to store parts between separate operations, and shortened the inspection time by consistently holding tolerances to ±.003 inch.

Part: Channel bars, 12' long, 3" wide
Lot size: 100
Machine Operations: (1) 18 holes drilled $\frac{11}{32}$", 1 hole drilled $\frac{13}{32}$"
(2) 11 holes drilled and tapped $\frac{1}{4}$-20, 1 hole drilled $\frac{13}{32}$"
(3) 3 holes drilled $\frac{13}{32}$"

AUTOMATIC CHANGING OF BORE SIZE

MACHINE: Horizontal-spindle precision bore
CONTROL: Four-axis positioning

Automatic tool changing can be accomplished with numerical control by a system in which the tool size is altered as desired (Fig. 7–5). A tape-controlled bore-sizing head-adjustment axis permits programmed boring of an infinite number of holes varying in size from a ¼-inch minimum to 4¼-inch maximum within a 1-inch diameter variation per tool. The cutting diameter of the tool may be programmed any number of times during the machining cycle.

Point-to-point tape control adjusts bore size. After machining each bore, the bar can be adjusted to eliminate tool-drawback lines, or it can be adjusted to a larger size and bore out. Tool offset controls for diameter and depth eliminate the need for tools to be accurately set to size. The boring-bar position is changed while the spindle is running.

By exploiting the tape-controlled bore-sizing head adjustment to its full extent, straight boring, step boring, feed and flycut facing, grooving, trepanning, and chamfering operations can be programmed in the most efficient sequence with the use of only one tool. Work handling and setup time are thereby substantially reduced. The fourth axis also reduces the number of tools and setting gages required to do a job. And, since only a few tools are sufficient to handle the entire range of possible bore sizes, the more elaborate multiple automatic tool changer is not required.

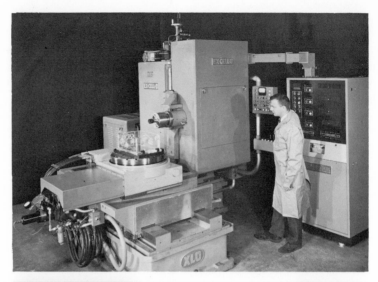

Fig. 7–5. Automatic changing of bore size: *top*—precision bore with experimental workpiece in place; *bottom*—experimental workpiece. (Courtesy of Ex-Cell-O Corporation.)

Typical Sequence of Operations. It took the following 85 steps to process the experimental workpiece shown on the Ex-Cell-O Model 992 machine tool in Fig. 7–5. Machine operations included facing, grooving, counterboring, and boring.

1. Slides move to work position; adjust tool to 3.250″ dia.
2. Facing three bosses.
3. Boring slide retracts.
4. Table indexes to position 2.
5. Slides move to work position; adjust tool to 2.250″ dia.
6. Facing rectangular boss.
7. Boring slide retracts.
8. Table indexes to position 3.
9. Slides move to work position; adjust tool to 3.250″ dia.
10. Facing one oblong and one round boss.
11. Boring slide retracts.
12. Slides move to work position; adjust tool to 3.062″ dia.
13. Counterbore 3.062″ dia. to 0.8125″ depth.
14. Head faces bottom of counterbore.
15. Adjust tool to 2.250″ dia.
16. Bore 2.250″ dia.
17. Head and boring slide retract.
18. Table indexes to position 1.
19. Boring slide moves to work position; adjust tool to 2.250″ dia.
20. Bore 2.250″ dia.
21. Head and boring slide retract, stop cycle; tool change (select tool 2).
22. Slides move to work position; adjust tool to 1.500″ dia.
23. Bore 1.500″ dia.
24. Head and boring slide retract.
25. Table indexes to position 3.
26. Slides move to work position; adjust tool to 2.250″ dia.
27. Counterbore 2.250″ dia. to 0.5625″ depth.
28. Head faces bottom of counterbore
29. Adjust tool to 1.500″ dia.
30. Bore 1.500″ dia.
31. Head and boring slide retract.
32. Table indexes to position 2.
33. Slides move to work position; adjust tool to 1.562″ dia.
34. Bore 1.562″ dia.
35. Head retracts tool.
36. Adjust tool to 2.500″ dia.
37. Adjust tool to counterbore 2.500″ dia. to 0.9375″ depth.
38. Head faces bottom of counterbore.
39. Facing between holes.
40. Head faces bottom of counterbore.
41. Bore 2.500″ dia.
42. Adjust tool to 1.562″ dia.
43. Bore 1.562″ dia.
44. Head and boring slide retract, stop cycle; tool change (select tool 3).
45. Table indexes to position 3.
46. Slides move to work position; adjust tool to 1.750″ dia.
47. Counterbore 1.750″ dia. to 1.000″ depth.
48. Head faces bottom of counterbore.
49. Adjust tool to 1.250″ dia.
50. Bore 1.250″ dia.
51. Head and boring slide retract.
52. Table indexes to position 4.
53. Slides move to work position; adjust tool to 1.000″ dia.

54. Bore 1.000″ dia.
55. Head and boring slide retract.
56. Slides move to work position; adjust tool to 1.000″ dia.
57. Bore 1.000″ dia.
58. Head and boring slide retract.
59. Slides move to work position, adjust tool to 1.000″ dia.
60. Bore 1.000″ dia.
61. Head and boring slide retract.
62. Slides move to work position; adjust tool to 1.000″ dia.
63. Bore 1.000″ dia.
64. Head and boring slide retract.
65. Table indexes to position 1.
66. Slides move to work position; adjust tool to 1.750″ dia.
67. Counterbore 1.750″ dia. to 1.000″ depth.
68. Head faces bottom of counterbore.
69. Adjust tool to 1.000″ dia.
70. Bore 1.000″ dia.
71. Head and boring slide retract;

stop cycle; tool change (select tool 4).
72. Boring slide moves to groove position.
73. Head grooves to 1.938″ dia.
74. Head and boring slide retract.
75. Table indexes to position 3.
76. Slides move to work position.
77. Head grooves 1.938″ dia.
78. Head and boring slide retract; stop cycle; tool change (select tool 5).
79. Slides move to work position; adjust tool to 0.375″ dia.
80. Bore. 0.375″ dia.
81. Head and boring slide retract.
82. Slides move to work position; adjust tool to 0.375″ dia.
83. Bore 0.375″ dia.
84. Head and boring slide retract.
85. Table indexes to position 1; stop cycle; tool change (select tool 1).

PRECISION MACHINING OF SMALL QUANTITIES

MACHINE: Precision toolroom
CONTROL: Contouring

In the modern toolroom and experimental shop, numerical control is proving of substantial assistance in the rapid and accurate production of small quantities of complex parts held to close tolerances. Features not practical or feasible with conventional machines are incorporated in numerical control machines because of the greater opportunity to utilize their merits. One versatile toolroom system, for example, performs highly accurate finish-boring, turning, grinding, and milling operations. Versatility is achieved with interchangeable operating heads and fixture units used singly or in combination.

A turning and boring head allows contouring of workpieces with the cutting speed maintained constant by varying the spindle speed to suit the change in peripheral diameter (Fig. 7–6).

Fig. 7–6. Precision machining with numerical control (Courtesy of Ex-Cell-O Corporation): *top*—contour turning of hemispheres; *bottom*—double setup for hemispheres. (*Continued on next page.*)

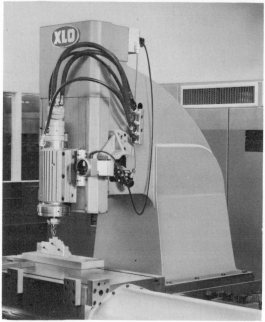

Fig. 7–6 (*continued*): *top*—contour boring of hemispheres; *bottom*—
three-axis contour milling.

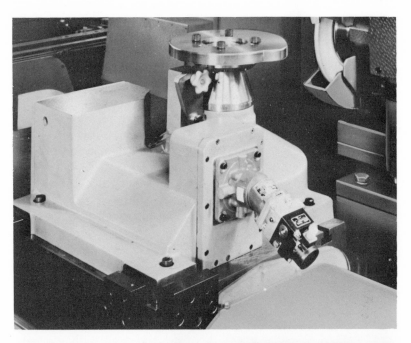

Fig. 7–6 (*concluded*): *top*—two-axis grinding of circular cams; *bottom* —two-axis grinding of templates.

A reciprocating grinding head incorporates a wheel dressing unit as an integral part operated as a periodic auxiliary function controlled by the tape. As a demonstration of accuracy, a male and female template ground on the system, when placed together over an illuminated sight gage, exhibit virtually no leakage of light over the complete length of the form.

A vertical milling unit (assuming three-axis control) allows vertical contour milling to very close tolerances.

PRECISION CONTOUR MILLING WITH INTERCHANGEABLE GRINDING

MACHINE: Contour milling
CONTROL: Contouring

"Space Age" accuracies are produced with a numerical control system utilizing a grinding head, complete with drive and reciprocating mechanism, which may be interchanged building-block style with the regular milling spindle and drive (Fig. 7–7). Tolerances of cutter path to programmed path are held to ±0.0001 inch. The control resolution is 0.000020 inch. Although primary usage of such equipment is for the nuclear and aerospace industries where ultraclose tolerances are essential, precision contour milling and grinding with numerical control are proving highly valuable in the die-sinking industries and elsewhere for accurate all-purpose requirements.

MACHINING COMPLEX SHAPES, I

MACHINE: Profiling
CONTROL: Contouring

A custom numerical control profiling machine solved production problems in making airfoil contours of steam-turbine buckets and jet-engine blades, certain types of drum cams, and other parts having intricate finished forms (Fig. 7–8). Workpieces are mounted between centers or chucked at one end, and the airfoils are machined completely at one setting. The principle of operation is that a milling cutter or a grinding wheel (depending upon the application) generates the radial contour and feeds for the length of the part, which may be either continuously rotated or indexed successively for each

Fig. 7–7. Interchangeable milling and grinding: *top*—precision milling spindle mounted on vertical slide; *bottom*—grinding attachment inter-changeable with milling spindle. (Courtesy of Ex-Cell-O Corporation.)

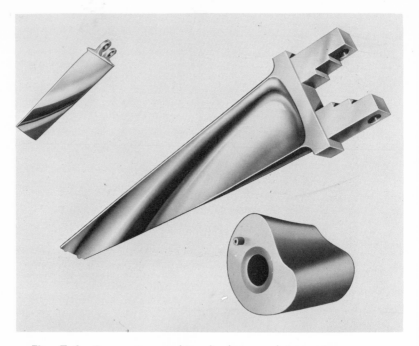

Fig. 7–8. Cams, steam-turbine buckets, and jet-engine compressor blades representing typical shapes produced. (Courtesy of Ex-Cell-O Corporation.)

cut. During a typical longitudinal milling operation, the work spindle is held stationary. The cutter, mounted on a precision spindle, is fed to depth, and the table is caused to feed for a distance equal to the length of the part. Cutting capacity is up to 60 inches in length and 11 inches in diameter. At the end of each stroke, the work spindle indexes by a small increment, and a cut is made on the return stroke. The cycle is repeated until the full 360 degrees are covered. Finish grinding can be accomplished with the work spindle continuously rotating.

MACHINING COMPLEX SHAPES, II

MACHINE: Profiling
CONTROL: Contouring

Numerical control profiling machines produce such parts as cams, templates, and airfoil contours of steam-turbine buckets and jet-engine blades. Parts having irregular and twisted forms can be machined without cams or special tooling. All the instructions for the machining operations are recorded on tape. The length of time required to make the tape is dependent upon the complexity of the part but is only a fraction of the time required to make the conventional tooling to do the same work on an operator-controlled machine.

In this machine, the position of the grinding wheel in relation to the work remains constant regardless of the dressing operation. This is true because the dresser does not move, but the wheel spindle moves toward the dresser, and the dressing operation has the effect of an automatic sizing device, restoring the wheel to its original position in relation to the work. Thus, the amount dressed off the wheel is compensated for automatically. The pause in the machining cycle, the withdrawal of the cross slide for the dressing operation, and the frequency of dressing are controlled by the tape.

The work is held either between centers or in some type of work-holding device (Fig. 7–9). During grinding and in rotary milling operations, one motor drives the work spindle and, through a train of gears, also drives the lead screw to give the desired feed per revolution. Longitudinal milling is an alternate method, whereby the work feeds lengthwise in relation to the milling cutter, then indexes and feeds back. Thus, feeding and indexing continue until the work is rotated 360 degrees. For this operation, a second motor drives the lead screw, and at the end of each stroke the first motor indexes the work spindle. All operations are under tape control.

PRECISION MILLING OF CAMS, I

MACHINE: Cam milling
CONTROL: Contouring

Contour milling operations on a variety of small precision components where production of consistently accurate complex forms is a requirement are done on a numerical control machine developed originally to improve production lead time and product quality on intricate three-dimensional fuel metering drum cams for jet engines (Fig. 7–10). In the initial application, machining time by the numerical control method

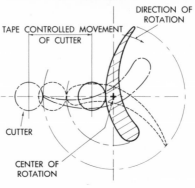

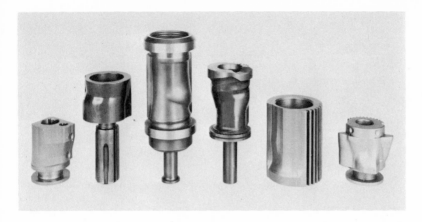

Fig. 7–10. Three-dimensional cams milled completely at one setting. (Courtesy of Ex-Cell-O Corporation.)

showed a 10:1 decrease from that of previous methods; tolerances were improved and became more consistent. Hand finishing was eliminated.

PRECISION MILLING OF CAMS, II

MACHINE: Cam milling
CONTROL: Contouring

The fuel-control systems of jet engines for planes and missiles require cams that are intricate and accurate. They are three-dimensional drum cams, usually not more than an inch and a half in diameter and about an inch long. The complete cam form is milled from solid cylindrical blanks in one continuous operation on a numerical control machine (Fig. 7–11).

The control system used with a cam-milling machine for this work has a pulse value of 50 millionths of an inch, which gives close control of machine movements, resulting in accurate cam forms.

To change to a new form, the operator instals the appro-

Fig. 7–9. Precision profiling: *top*—milling of jet-engine compressor blades; *center*—roller-type steady rests prevent deflection of the steam-turbine bucket during milling; *bottom*—diagram of cutter and section through airfoil showing how the cutter moving in one path and one plane generates the desired form. (Courtesy of Ex-Cell-O Corporation.)

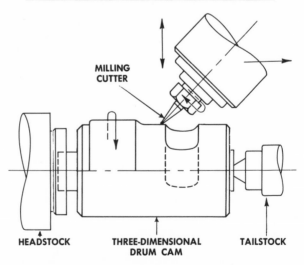

MILLING
CUTTER

HEADSTOCK   THREE-DIMENSIONAL   TAILSTOCK
DRUM CAM

priate tape in the control cabinet, loads the part, sets the slides and work head to the start position, and initiates the automatic cycle. The cutter takes repeated light cuts, completely finishing the cam form from blank to finish size in one operation.

PRECISION GRINDING OF CAMS

MACHINE: Cam grinding
CONTROL: Contouring

Ultraprecision flat circular cams are produced automatically, without the need for manual layout or the preparation of a master numerical control. A two-axis precision cam-grinding machine has a scribing attachment for rough scribing the cam blanks using the control tape. Use of the attachment ensures the preparation of a rapid and precise cam layout on blanks prior to the roughing-out process. This automatic layout is a rapid operation and is performed in a fraction of the time necessitated by conventional hand methods. When the cam is ready for finish grinding, the same control tape is used.

Cams may be produced in any quantity without the need for a master, the necessary control information being permanently stored on the tape. This feature considerably reduces lead time from that of conventional methods necessitating much handwork, and, at the same time, the user is assured of consistent accuracy in the cam form (Fig. 7–12).

USING NUMERICAL CONTROL WITH SEPARATE MACHINING OPERATIONS

MACHINE: Small machining center
CONTROL: Contouring

When part requirements are cyclical, it is particularly convenient if space-consuming tooling does not have to be stored

---

Fig. 7–11. Precision milling of cams: *top*—milling cutter path controlled by tape, with work-spindle rotation geared to longitudinal traverse; *lower left*—typical fuel-control cams the forms for which are completely milled from a cylindrical blank in one machining operation; *lower right*—schematic drawing showing directions of movement of major machine components in milling a three-dimensional drum cam. (Courtesy of Ex-Cell-O Corporation.)

Fig. 7–12. Grinding of cams: *left*—vertical oscillation of the grinding wheel results in a straight cam face; *right*—capacity of same machine includes cams of from 3½ to 14 inches in diameter and up to ¾ inch thick. (Courtesy of Ex-Cell-O Corporation.)

Fig. 7–13. Numerical control machining in two separate operations: *top*—part orientation for operation number 1; *bottom*—part orientation for operation number 2. (Courtesy of Michigan Tool Company.)

during the in-between periods. A part typical of those produced by a company manufacturing packaging machinery is required in lots of 225 to 550 periodically (approximately six months apart). The part is run across a tape-controlled tool-changing machine (machining center of the smaller type) in two separate operations, which include deburring and stamping of the part number on each piece (Fig. 7–13); the last two are done simultaneously while the machine is running. The last operation is inspection on a sample-lot basis.

Records show 21.5 hours for programming and necessary paperwork prior to making the tape. Following is a breakdown of allowed times:

Operation number	1	2
Machine cycle	5.96 min.	3.60 min.
Load and unload piece	0.74 min.	0.94 min.

Each operation takes approximately 1.0 hour for machine setup and 1.5 hours for tool setup. There were *no* tooling costs for jigs and fixtures; standard equipment was used.

Importantly, the programming time was expended only once. Yet, no special tools are stored except for the tape, which is available at a cost of $0.53.

MACHINING CENTER QUALITY CONTROL

MACHINE: Machining center
CONTROL: Contouring

The introduction of numerical control has prompted changes in quality-control procedures in many companies.

One company manufactures a part (Freon housing) that requires end milling, profile milling, drilling, tapping, reaming, boring, counterboring, cutting with a form tool, countersinking, and chamfering (Fig. 7–14).

With conventional techniques, the part requires an average of 13 hours to make in lots of 30 to 50. The intricacy of operations is a source of frequent error that results in serious time loss when discovered (usually at assembly). Therefore, 2 extra parts per lot are made to serve as spares.

On a numerical control machining center, the same part is made in two setups in 2.4 hours. In addition to the considerable

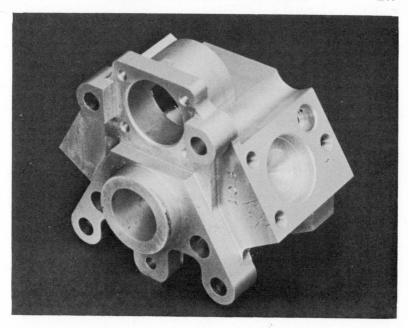

Fig. 7–14. Part requiring end milling, profile milling, drilling, tapping, reaming, boring, counterboring, cutting with a form tool, countersinking, and chamfering. (Courtesy of Cadillac Gage Company.)

time savings in direct machining, the quality control is vastly easier. Quality-control personnel make a comprehensive check of the first part of each lot. If it passes, the tape is stamped by them and subsequent parts in that lot are given a much quicker review in which only the most critical dimensions are checked.

REDUCED HANDLING OF LARGE WORKPIECES

MACHINE: Horizontal single-spindle universal tool
CONTROL: Positioning

Machine-tool companies are among the most prominent users of numerical control machines. In the first place, machine-tool parts tend to be in families so that, except for dimensions, the configurations are similar. Tape preparation can, therefore, be facilitated. Secondly, machine tools utilize exceptionally large components, making it desirable to minimize handling.

One such component is a cast-iron support table (35½ by 20 by 16 inches) which is box-milled on two ends, pad-milled

Fig. 7–15. Support table mounted on machine. (Courtesy of Associated Machine Company.)

on two sides, and drilled and tapped (16 holes ¼-120). The part is revolved to machine all four sides (Fig. 7–15).

Such a large part is easily completed on a tape-controlled machine with a 40- by 40-inch X-Y range. Conventional machining would necessitate almost prohibitive multiple handling.

MACHINING VERSATILITY WITH NUMERICAL CONTROL

MACHINE: Single-spindle universal tool
CONTROL: Positioning

Numerical control machines have made it possible for numerous companies to expand their capacity dramatically with relatively little added investment, owing to the versatility of numerical control equipment.

A manufacturer of precision cameras with a business volume that had been practically level for three years had a rapid spurt in business from $2 million to $7 million in 18 months. One of the company's typical parts is a camera case made

NUMERICAL CONTROL APPLICATIONS

from an aluminum casting (Fig. 7–16). The part has an overall finished size of 8.312 by 7.281 by 7.500 inches and is machined on all sides—four sides with a numerical control setup and the other two by conventional milling. Operations include end milling, face milling, drilling, tapping, reaming, and boring in a wide range of sizes with tolerances exemplified by that of the rail slot of + 0.001 − 0.000. Eight tapes are used to complete the job.

The company acquired 13 numerical control universal tools, which quickly put it into a position to accommodate the increased activity. Yet, the investment was for universal-type equipment not uniquely tied to the parts made at the time.

VERSATILITY OF SINGLE-SPINDLE UNIVERSAL TOOL

MACHINE: Single-spindle universal tool
CONTROL: Positioning

A carbon-steel front-roller support (20 by 34 by 11 inches) for a motor-winding machine requires 4 pads to be milled and 12 holes to be drilled, with 4 of the holes to be tapped and 8 bored. The bearing support is counterbored from both sides 4 inches deep. Tolerance on the diameter is 0.0002 inch, and on the centerline of holes 0.001 inch. The depth of holes is held to 0.002 inch. The whole job is done on a single-spindle numerical control machine (Fig. 7–17). The size of the part and the nature of the operations would require a much greater investment in conventional machines to do the same job.

USE OF TAPE-CONTROLLED ROTARY WORKTABLE

MACHINE: Three-axis turret drill with numerical control rotary table
CONTROL: Positioning

Numerical controls facilitate equipment features that would not be particularly useful in conventional machines. For example, a tape-controlled turret drilling machine can handle a much broader range of work when it is equipped with a rotary table that is also tape-controlled (Fig. 7–18). A part mounted in such a machine can be oriented in virtually any position with very little operator handling and tooling. The operator's job is

Fig. 7–16. Precision camera case mounted on machine table. (Courtesy of Photo-Sonics, Inc.)

Fig. 7–17. Part mounted vertically for multiple operations by single-spindle universal tool. (Courtesy of Woodrow Brixius, Inc.)

Fig. 7–18. Three-axis numerical control turret drill equipped with numerical control rotary table. (Courtesy of Burgmaster Corporation.)

made much easier. The machine becomes quite universal in its capacity to do multiple operations with minimum setup and changeovers.

TAPE-CONTROLLED TURRET MACHINING

MACHINE: Three-axis turret drill
CONTROL: Positioning

Not only did a major machine-tool builder-user achieve a 25 per cent reduction in the time required to machine a part with its numerically controlled turret drilling, tapping, boring, and milling machine, but also $1,200 were saved in tooling costs (Fig. 7–19). In addition, the control combined with the machine resulted in substantially more uniform parts in the 24-piece-lot runs. This made possible additional savings in reduced inspection time and emphasizes the advantage of tape control for small-lot runs.

Fig. 7–19. Extreme simplicity of tooling for job setup resulted in $1,200 tooling savings: *top*—parts being drilled, tapped, bored, and milled in one setup, saving considerable handling time; *bottom*—the finished part after machining by numerical control. (Courtesy of Ex-Cell-O Corporation.)

TURRET DRILLING: COST-PER-PART COMPARISON

MACHINE: Turret drill
CONTROL: Two-axis positioning

An instrument part (Fig. 7–20) with the following specifications was produced both on conventional machines and on a numerically controlled turret drill:

Material	2024 ST 3 aluminum alloy
Number of piece parts	40
Part dimensions	10″ × 7″ × 0.375″
Number of holes	22
Hole diameters	9 at 0.172″ diameter
	4 at 0.218″ diameter
	4 at 0.270″ diameter
	5 at 0.500″ diameter

The table below lists the time and cost figures for producing this part by the two methods. Labor costs vary in the area where the manufacturer is located, but the figures here are considered representative for the area. All labor costs are burdened (overhead, etc.).

Cost Comparison

Operation	Labor Cost per Hour	Conventional Time (hrs)	Conventional Cost ($)	Numerical Control Time (hrs)	Numerical Control Cost ($)
Conventional planning	$9.20	2.0	18.40		
Numerical control programming[1]	9.20			2.0	18.40
Tool design	9.20	20.0	184.00	2.0[2]	18.40
Tool fabrication[3]	7.50	24.5	183.75	6.0	45.00
Tool material			45.00		10.00
Tape preparation (clerical)	4.60			1.25	5.75
Setup	7.50	0.5	3.75	1.00	7.50
Total initial time and costs (planning, tooling, setup)		47.0	434.90	12.25	105.05
Initial cost/part			10.87		2.13

Fig. 7–20. An application of turret drilling: *top*—instrument panel; *bottom*—panel mounted on machine table with simple clamping. (Courtesy of Remex Electronics.)

Cost Comparison *(continued)*

Operation	Labor Cost per Hour	Conventional		Numerical Control	
		Time (hrs)	Cost ($)	Time (hrs)	Cost ($)
Operator/part	7.50	0.42	3.15	0.005[4, 5]	0.037
Machine operation/part[6]				0.067	0.502
Operating costs (operator, machine)			3.15		0.54
Total cost/part			14.02		2.67

[1] The control system is incremental so that the programming can be done directly from the blueprint. It is not necessary to convert these dimensions to Cartesian coordinates.

[2] The numerical control tooling consists of a simple holding fixture with no bushing plate or bushings.

[3] Tooling costs are average job shop labor and overhead costs.

[4] Operator need only load and unload parts. He can spend operating time deburring, etc.

[5] It should be noted that the actual operating time per part is three minutes, fifteen seconds. In the case presented, the part is drilled with spiral point drills. Center drilling is not necessary. For cases where center drilling is desired, the operating time would increase slightly. The relative costs, however, can be seen to still heavily favor the numerically controlled tool. The example shown is for a two-axis control system. Three axes of control provide substantial additional savings.

[6] Machine-operation cost includes amortization of machine tool and control system at one-shift usage, 100 per cent depreciation in ten years, 6 per cent interest on invested funds, maintenance cost, spare parts, floor space, power, property taxes, and insurance.

8

Automation and the Future

Science and Technology

Except for a few relatively short periods in world history during which scientific study was conducted as a pure quest for knowledge, most technical research and development have been generated out of necessity. However, the twentieth century has seen a veritable explosion of activity in every conceivable field, running the gamut from pure theory to the most practical application and implementation. When one considers that the body of knowledge doubles every fifteen years, it is clear the pace is truly fantastic. The constantly expanding horizons of technical and scientific knowledge represent one of the most compelling forces pushing the world toward increasingly intensified utilization of automation. What started out not too long ago as spotty hit-and-miss attempts to decrease the manual effort required to produce articles in a few industries has grown into a high-gear activity in which products and services are made in quality and quantity that were not even considered a few years ago. In turn, some of these same products and services are promoting technological advancement by making possible instruments and techniques that facilitate scientific inquiry. Therefore, advances in technology make possible advances in automation, which make possible further advances in technology, and so on. This feeding on each other is characteristic of the automation-technology relationship and is a powerful contributor helping the most optimistic of early predictions regarding automation approach reality.

Population

In the first 1600 years of the Christian era, the earth's population doubled to 500 million people. It doubled again to 1 billion in the next 250 years and once more to 2 billion in the following 80 years. Today, some 40 years later, the population is over 3 billion. By the turn of the century, it will double once more.

The impact of this population explosion on the need for greater productivity is overwhelming. It has been pointed out that in food production alone, nearly 10,000 years of agricultural progress must be matched in this generation to prevent the population from outstripping the food supply. Further, the mass of goods and services that will be required by this population staggers the imagination. Can there be any doubt that automation is the key to meeting this challenge?

The growing demands on our ability to produce goods are a result not only of the over-all increase in population but also of the fact that the population is becoming largest in that age group (20 to 29) which purchases and consumes the greatest amount of goods. It is in this age group that most people have new discretionary purchasing power, marry, set up new households, move, and accelerate their economic progress.

The burgeoning population and its makeup represent enormous demands for those things that man must make, but can make only with means that will expand his productivity far beyond present levels. Hence, the turn to automation.

Communications

The rapid growth in technology has also made the world smaller through new and faster transportation and communications methods. The significant result has been that there is now an almost universal awareness of how everybody else in the world lives. The sociological and economic implications of this awareness are far-reaching. As millions of people in remote, economically depressed regions become familiar with elements of living standards in better-endowed areas, pressures mount for these people to obtain some of the same things. Over the past two decades, the standard of living worldwide has risen to new levels. This rising standard has enabled more people to acquire the necessities of life and turn their attention to upgrading the things they do and the things they buy—whether home furnishings, automobiles, luxury items, or leisure-time activities. It is evident that better communications shrink the world and help bring about rising living standards over ever widening areas, simultaneously increasing the demand for more and more goods and services.

Competition

Not only the United States but all nations are confronted with the necessity to satisfy the demands of an increasing population, the desire to improve standards of living, and the challenge of constantly expanding frontiers of technical and scientific knowledge. Several nations have an interest in the international race for technical and scientific supremacy and the ultimate capacity to produce and deliver nuclear armaments and control outer space. Finally, all nations are interested in their own economic well-being. This is not only a function of their political posture or alignment; it is determined to a larger extent by their abilities to cope with international competition in trade. They must reduce costs to expand markets or be increasingly at the mercy of other nations. "Cheap labor" no longer offers the competitive edge in the manufacture of many products.

It is this competition that should be of particular concern to the managers of industry. A company's competition is no longer limited to the neighbor who happens to make a similar product. We can observe that modern communications media (used here in the broadest sense to include transportation and distribution) reduce the effects of distance alone in precluding competition. The tendency, in fact, is that companies are becoming less local so as not to restrict their markets. What used to be national companies that happened occasionally to do international business are now international companies that happen to do business in particular countries. This magnifies the competition among them and points to productivity— hence, automation—as the major factor determining which will succeed.

INDUSTRY TRENDS: NEED FOR FLEXIBILITY

Thus far, the discussion has centered on the trends that emphasize the importance of greater productivity. It was suggested that this productivity is attainable via increased utilization of automation. We must be careful, however, not to think of productivity and automation as being entirely synonymous.

Automation signifies much broader implications not the least of which is flexibility. This aspect of the forces creating pressures for greater use of automation becomes particularly apparent in analyzing what is increasingly happening on the company level. In recent years, additional demands have been made which are not adequately satisfied by the tools of mass production as we know them.

Shorter Product Life

The general trend in almost every consumer product is toward a higher rate of obsolescence, that is, shorter product life. Automobiles, for example, are not intended to be useful for as long a period as the earlier cars. Along with this trend toward shorter product life, automobile manufacturers provide replacement parts for only a short time (three years for many parts). Additionally, of course, the demand for replacement parts is quite small compared to the quantities involved in the initial product. Consequently, the extraordinary investment required for superefficient special-purpose machines of mass production becomes all the more prohibitive because of the high obsolescence factor involved.

Cyclic Demand

There are some products which do not necessarily change in design requirement substantially with time but which have cyclic demand. For example, farm equipment of most types generally sells in early spring and is actually traded in late summer or early fall. This type of product, because of both the quantities involved and its constant nature, is not readily amenable to the efficient application of the typical mass-production special-purpose machines. This fact is confirmed by the extensive use of universal machines in plants where small production runs resulting from cyclic demand are the rule.

Engineering Changes

A trend that has become quite apparent in recent years is the acceptance of engineering changes after the product has been released for manufacture. Perhaps the major influencing factor in this direction has been the military organizations, which have let contracts involving large complex systems, almost always requiring the integration of engineering and manufacturing contributions of a number of companies. Although the changes are most often justified, the effect on manufacturing represents a distinct change from what was the case not many years past. Once it was almost impossible for engineering changes to be approved once tooling for a product was completed and in use. Yet, it is not considered unusual that a complex fire-control system had over 30,000 changes before the first completed system was out of the manufacturing plant.

Engineering changes have also been made acceptable by the needs of prototype manufacture and low-quantity first lines. For instance, certain jet-aircraft manufacturers can ill afford to prohibit engineering changes until considerable experience is obtained with their product. The situation is similar for many other products which are considerably less complex.

What do these engineering changes mean? Obviously, they play havoc with the special-purpose machines of mass production. Typical is what happened to a major manufacturer of turbine engines. Upon delivery of a special machine built to fabricate a particular part to certain dimensions, the machine had to be routed to a warehouse for storage and future cannibalization of components because the part in question had already been changed dimensionally in design.

Buyers' Market

The return of the buyers' market has probably struck the severest blow to special-purpose machines. For example, it was not unusual in 1960 for an automobile brand to be available in thirty models. If all the possible configurations, accessories, color combinations, body styles, and the like of automotive equipment available in 1967 were listed, the number would indeed be staggering. And this trend is not unique to the automobile industry. There are many product lines where a large number of varying specifications are received from customers for the same product. This means merely that the machinery available to provide this equipment must have the capacity to handle the variations.

It is not suggested that the aforementioned trends will cause the mass-production machines to disappear. Certainly that will not happen for some time to come. However, the trends do indicate a change of emphasis. Specifically, although large-quantity production of fixed interchangeable products requiring very special inflexible machines has been the pattern until very recently, now the need is for extremely flexible machinery able to cope with small-lot production of many variations of a product.

It is clear that industry will have to be flexible itself to keep up with the ever increasing demands for products in all shapes and forms at costs the consumer has become accustomed to through mass production. The challenge, in fact, seems to be that the customer must be satisfied—and he will not be until the ultimate is reached. That is, he shall want to specify what he wants and the production machinery, besides anticipating his wants, will have to provide the desired item with what amounts to individual attention at costs comparable to those of making the item in very large quantities. Fortunately, most of industry is coming to this realization and steps in the right direction are being taken.

Index

233